Modern Applications of DNA Amplification Techniques

Problems and New Tools

Modern Applications of DNA Amplification Techniques
Problems and New Tools

Edited by

Dirk Lassner

Institute of Clinical Chemistry and Pathobiochemistry
University of Leipzig
Leipzig, Germany

Barbara Pustowoit

Institute of Virology
University of Leipzig
Leipzig, Germany

and

Arndt Rolfs

Clinic and Policlinic of Neurology
University of Rostock
Rostock, Germany

Plenum Press • New York and London

Library of Congress Cataloging-in-Publication Data

Modern applications of DNA amplification techniques : problems and new
tools / edited by Dirk Lassner, Barbara Pustowoit, and Arndt Rolfs.
    p.   cm.
    "Proceedings of the Augustusburg Conference of Advanced Science on
Problems of Quantitation of Nucleic Acids by Amplification
Techniques, held September 23-26, 1996, in Augustusburg, Germany"-
-T.p. verso.
    Includes bibliographical references and index.
    ISBN 0-306-45801-2
    1. Polymerase chain reaction--Congresses. 2. DNA--Synthesis-
-Congresses.   I. Lassner, Dirk. II. Pustowoit, Barbara.
III. Rolfs, Arndt, 1959-   . IV. Augustusburg Conference of
Advanced Science on Problems of Quantitation of Nucleic Acids by
Amplification Techniques (1996)
    [DNLM: 1. DNA--analysis--congresses. 2. Polymerase Chain
Reaction--methods--congresses. 3. Polymerase Chain Reaction-
-standards--congresses. 4. Virus Diseases--diagnosis--congresses.
QU 58.5 M689 1997]
QP606.D46M63 1997
572.8'645--dc21   97-41939
                                                                 CIP

QP
606
.D46
M63
1997

Proceedings of the Augustusburg Conference of Advanced Science on Problems of Quantitation
of Nucleic Acids by Amplification Techniques, held September 23 – 26, 1996,
in Augustusburg, Germany

ISBN 0-306-45801-2

PREFACE

In the ten years since the first publication on PCR (Saiki et al., 1985), this *in vitro* method of nucleic acid replication and modification has grown to rival in popularity traditional microbiological, genetical und technical procedures for cloning, sequencing, gene detecting and related procedures. To date the PCR literature has emphasized six main areas of application: genetic mapping, detection of mutations, genetic polymorphism, transcriptional splicing and regulation, molecular virology and quantitative procedures. The overwhelming focus of quantification of DNA or RNA by PCR has been on human microbiology and oncological problems. The exquisite sensitivity of PCR gives this method the ability to detect extremely rare DNAs, mRNAs, mRNAs in small numbers of cells or in small amounts of tissue, and mRNAs expressed in mixed-cell populations. However, the exact and accurate quantification of specific nucleic acids in biological samples is in spite of numerous publications in that field still a general problem: during the PCR process, an unknown initial number of target sequences are used as a template from which a large quantity of specific product can be obtained. Although the amount of product formed is easy to determine, it is difficult to deduce the initial copy number of the target molecule because the efficiency of the PCR is largely unknown. Several experimental factors may affect the efficiency of PCR amplification: the sequence being amplified, the length of the target, the sequence of the primers and potential impurities in the sample. There are additional, as yet unknown, often subtle factors that also affect the amplification efficiency. Some of the earliest experiments designed to quantify nucleic acid levels involve the measurement of external standards. Nevertheless, in an attempt to correct for tube-to-tube variations in amplification efficiency, an internal standard seems to be the most accurate procedure at the moment.

As a sign for still existing difficulties in the methodological field of PCR quantification there are even some highly sophisticated mathematical papers dealing with that problem (e.g. Nedelman et al. 1992).

To give an actual overview of the state of the art in PCR quantification the idea for a conference on the Augustusburg near Dresden was born during a discussion between the editors and other scientists. The title of the conference was "Modern application of DNA amplification - problems and new tools" and the aim was to present a thorough discussion of problems in that field. However, there is now doubt that due to the vast amount of available information about PCR and the rapid development of new methods, the editors and organizers may have inadvertently omitted some important and exciting information. The results of all the investigations in the rapidly growing field of PCR technology give a lot of answers to actual questions. But with each new answer, more questions arise.

The editors would like to thank Joanna Lawrence and Robert Wheeler from Plenum Press whose infinite patience, encouragement and support have been invaluable. The editors are also grateful to Ms. Katrin Schmidt/Rostock for her patient assistance in all stages of this undertaking.

We thank all authors for their contributions and for their care in compiling useful practical accounts of their specialist areas of PCR technology. The editors hope that the reader will find that the book represents an informative and useful manual to continue the list of good PCR books.

Dirk Lassner
Barbara Pustowoit
Arndt Rolfs

CONTENTS

PCR Quantification and Technical Aspects

PCR Quantification of Infectious Agents

PCR Quantification and Technical Aspects

MULTIPLE COMPETITORS FOR SINGLE-TUBE QUANTIFI-CATION OF HIV-1 DNA

Tanya Vener[1], Malin Stark[1], Jan Albert[2], Mathias Uhlén[1] and Joakim Lundeberg[1]

[1]Department of Biochemistry and Biotechnology, KTH, Royal Institute of Technology, Teknikringen 30, 100 44 Stockholm. [2]Department of Clinical Virology, Swedish Institute for Infectious Disease Control, Karolinska Institute, 105 21 Stockholm, Sweden

INTRODUCTION

The polymerase chain reaction (PCR) has drastically changed the diagnostic possibilities to detect and analyse diseases caused by pathogens, such as viruses and parasites[1]. Furthermore, the use of PCR for quantitative analysis has become a key technique in disease prognosis. However, large scale routine quantification of clinical samples has been hampered by the amount of labor needed for accurate quantification. Analysis with limiting dilution and Poisson distribution gives a correct prediction but is labor intensive[2]. Instead, several methods based on co-amplification schemes have been described[3-7] involving mixing varying amounts of competitor with a constant amount of target. Most of these methods have the disadvantage that they require multiple tubes per analysis to assess the titration inflection point.

One important clinical area related to PCR quantification systems is the analysis of human immunodeficiency virus type 1 (HIV-1) during treatment with anti-viral drugs such as AZT, ddC and ddI[8] and it has been shown that direct measurement in viral load can be correlated to the effect of different treatment regimes.

Here, we describe an alternative for viral load analysis of HIV-1 in samples using multiple competitors in a semi-nested co-amplification procedure. The method has been designed to allow routine analysis in a non-isotopic format and to yield information on the amount of viral HIV-1 DNA in peripheral blood mononuclear cells (PBMC). The competitor DNAs have been designed to contain the same LTR primer binding sequences as the wild type DNA, but with different internal sequences and length. Discrimination between the wild type DNA and the four competitors has been performed by using one fluorescently labelled inner PCR primer followed by fragment analysis using standard automated sequencers. A calibration curve can be established using the peak area of the four competitors for accurate determination of target

Modern Applications of DNA Amplification Techniques
Edited by Lassner *et al.*, Plenum Press, New York, 1997

amounts with minimal tube-to-tube variations.

MATERIALS AND METHODS

Sample material

The model system described here used HIV-1$_{MN}$ infected peripheral blood mononuclear cells (PBMC) diluted in uninfected PBMC to contain various numbers of proviral HIV-1 copies. PBMC were isolated by Ficoll-Paque density centrifugation and lysed without prior cultivation in PCR lysis buffer (10 mM Tris-HCl pH8.3, 1 mM EDTA, 0.5% NP40, 0.5% Tween 20 and 300 µg/ml Proteinase K) at a concentration of 10^6 cells/100 µl as previously described[9]. Crude cell lysates were used directly for PCR amplification.

Construction of competitor DNAs

HIV-1 competitor DNAs were constructed by linker assembly into plasmid DNA as described previously by Vener *et al*[10]. An additional competitor, no.4 (136 bp fragment), was constructed by insertion and blunt-end ligation of a HaJo linker (36-mer, 5'GGGAACACCAT GAACACCACCATGACCCG-3' and 3'-TCGACCCTTGTGGTGGTACTTGT GGTGGTACTGGGCCTAG-5') at the *EcoRV* site of the previously published construct no.210. The competitors were subsequently serially diluted in 10 mM Tris-HCl pH 8.3 and 10 ng/µl yeast tRNA (Boehringer Mannheim, Mannheim, Germany).

Limiting dilution

The concentrations of competitors and HIV-1$_{MN}$ cell lysates (both of stock and diluted lysates) were determined using limiting dilution and nested PCR as described previously [2]. In short, the materials were diluted in five-fold steps and at least ten PCR determinations were performed on each dilution. The copy numbers were calculated by the Poisson distribution formula[2] (i.e. one starting copy corresponds to a dilution step in which 63% of the samples are positive). The basis for the analysis is the ability of the nested PCR to reproducibly detect single HIV-1 molecules.

Polymerase Chain Reaction

PCR primers annealing in the 3'- LTR region of HIV-1 were synthesised according to the manufacturer's recommendations (Pharmacia, Uppsala, Sweden). One of the inner primers (sense) was biotinylated and the other inner primer (antisense) was dye-labelled for fluorescence-based detection. PR primer sequences were JA159, 5'CAGCTGCTTTTTGCCTGTAC-3' (outer, 432-452); JA160F, 5'-FITC-CTGCTTTTTGCCTGTACTGGGTCTC-3' (inner, 435-460); JA161B 5'-biotin-AAGCACTCAAGGCAAGCTTTATTGA-3' (inner, 524-499); JA162, 5'-AGCACTCAAGGCAAGCTTTA-3' (outer, 528-508). Positions are given relative to MN[11] strain of HIV-1. Both inner and outer PCR were carried out in 50µl containing 5µl of 10-fold PCR buffer (100mM Tris-HCl, pH 8.3 at 25°C, 500mM KCl, 25mM MgCl$_2$) and 200µM of each deoxynucleotide triphosphate (dNTP) with 5pmol of each primer and 1 unit

Taq polymerase (Perkin-Elmer, Cetus, USA). The temperature profile consisted of denaturation, 94°C for 5 min, linked to the cycle program: 96°C for 30 sec; 50°C (outer) or 60°C (inner) for 30 sec; 72°C for 30 sec. 5µl of outer PCR product was transferred to the inner PCR. Both outer and inner PCR used 32 cycles (GeneAmp 9600, Perkin Elmer, Cetus, USA). Multiple negative controls were included in each PCR run. For quantification by competitive PCR, the four competitors were premixed in different ratios prior to amplification. Ten microliter competitor mixtures were added together with 10µl of sample to the reaction tube.

Fragment analysis

One microliter of the resulting PCR products was mixed with 9 µl of deionized formamide (100%), containing Dextran Blue 2000 (Pharmacia Biotech, Uppsala, Sweden), heat-denatured for 5 minutes at 95°C, immediately chilled on ice and loaded on a 6% polyacrylamide gel (Ready Mix, Pharmacia) and electrophoresed on an automated laser fluorescent sequencer [ALF] (Pharmacia). A 50bp fluorescent-marked ladder (50-500) was used as a size standard. Quantification and interpretations of the raw data output was facilitated by using Fragment Manager software (Pharmacia).

RESULTS

In this study we present an optimised method for the quantification of HIV-1 DNA using a PCR strategy with multiple competitors. The method is schematically outlined in figure 1. The method is based on the co-amplification of the sample DNA with four competitor DNAs which can be discriminated by length. The PCR primers anneal to the 3'-LTR region of HIV-1 and to equivalent LTR linker sequences in the competitor DNA. The resulting PCR products are analysed by electrophoresis using the internal competitors to create a standard curve employed to quantify the amount of target.

Construction of competitors.

Four different competitors were assembled into a pGEMz vector using synthetic linkers. Upstream and downstream linkers containing HIV-1 specific sequences of the LTR region were inserted into the multiple cloning site of the pGEMzpA vector, resulting in competitor no.1. To generate competitors of different lengths two extension linkers (*EcoRV* or *Lac* operator linker) were ligated into the first HIV-1 LTR competitor construct, resulting in competitors no.2 and no.3, respectively, as described previously[10]. A fourth competitor was constructed by insertion of HaJo linker (36bp duplex) at *EcoRV* rare site of construct no.2, resulting in a 136bp PCR fragment.

Quantification of HIV-1 DNA

In an initial study various amounts of HIV-1MN DNA were quantified using a similar strategy[10]. A premix of fixed amounts of three competitor DNAs of different length was used in a semi-nested competitive PCR. The corresponding lengths of inner PCR products varied over a short interval: competitors no. 1-3; 89 bp, 100 bp and 125 bp, respectively. Amplification of the target, HIV-1, resulted in a product of 114 bp, thus within the range covered by the competitors. The narrow size interval between the different products was deliberately chosen to minimise amplification differences, while still enabling baseline separation. The previous study showed that high quality data can be achieved enabling a rough estimation of the number of targets by a simple analysis of the raw data from chromatograms, irrespective of competitor configuration.

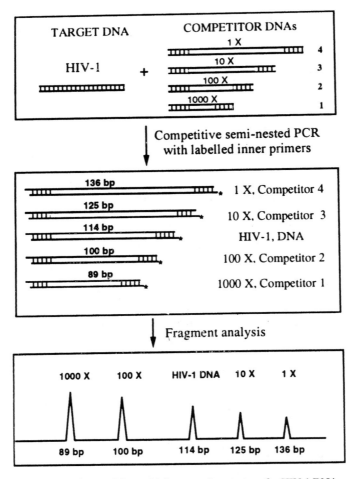

Figure 1. A schematic illustration of the multiple competitor strategy for HIV-1 DNA quantification.

Improvements for quantification of HIV-1 DNA

To improve the single tube quantification approach, a fourth competitor (136 bp fragment) was constructed on the basis of DNA construct no. 2. This allows for a more reliable estimation of the amount of HIV-1 DNA, since the calibration curve contains an additional standard. The fourth competitor can either be used to expand the dynamic range or to allow for a more exact quantification by making more narrow competitor dilution series. In this context, different configurations of the four competitors were analysed. An important consideration is the limitation in the detection step (i.e. the instrument detection range), since chromatographic „cut" peaks can not be reliably used in the construction of the internal standard curve. Figure 2 depicts the two alternatives: increased dynamic range or more exact quantification. Panel a shows a 5-fold configuration of competitors 10:50:250:1,250 and a HIV-1 DNA target corresponding to 500 starting copies. Here competitor no.4 can not be used in the construction of a standard curve due to the „cut" peak. However, by a 5-fold dilution of the same PCR product (panel b) competitor no.4 can be used, while competitor no.1 has a too low signal to be detected. However, both panels can be used for a direct estimation of HIV-1 DNA quantities by comparing with the competitor peak areas. To achieve a more exact estimation, a more narrow dilution series can be employed as depicted in panels c and d. Panel c shows a 2-fold configuration of competitors 10:20:40:80 and HIV-1 DNA target corresponding to 10 starting copies. Actually, panel d shows an example in which all 4 peak areas are used to create a standard curve, which was generated by a 5-fold dilution of the PCR product analysed in panel c.

DISCUSSION

We present a single tube assay for reliable quantification of HIV-1 DNA. The assay can be useful to monitor disease progression in HIV-1 infected individuals. In most strategies to quantify nucleic acids by PCR it has been shown that it is essential to include an internal control to correlate for amplification differences, since the presence of PCR inhibitors in individual samples may otherwise influence the results. The novelty with this approach is that several competitors are included into the same tube and that the PCR continues until the reaction is saturated, i.e. until the plateau is reached. This avoids the need to be in a narrow exponential phase of amplification for proper evaluation. Importantly, this nested PCR approach ensures high sensitivity as single-molecule detection can be achieved which is of great importance for precise quantification of HIV-1 DNA. Direct measurement of viral load will most likely give a more rapid and reliable answer than other indirect measures of virus activity[7].

The main benefit of using multiple competitors for quantification is that it eliminates the need to use multi-tube analysis to assess the inflection point in competitive titration experiments which previously has been the dominating approach, while still enabling the use of the same primers for amplification of the sample and the internal controls. Thus, this single tube analysis increases throughput drastically without effecting the robustness. Furthermore, a more accurate analysis or a wider dynamic range can be achieved by further increasing the number of competitors. From our results it appears that competitors consisting of a heterologous DNA-fragment flanked by cloned linkers with the primer annealing sites can be used as long as the competitor lengths do not differ significantly from the native target sequence length. We have used fragment analysis on automated laser fluorescent electrophoresis apparatus in order to discriminate between the different competitors. The evaluation can easily be performed within a few hours and the analysis software gives a precise determination of peak areas.

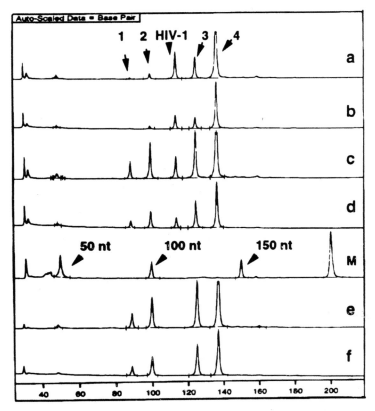

Figure 2. The fragment analysis results with various competitor configurations and amounts of HIV-1 target. Panel a and b corresponds to a 5-fold competitor configuration (10:50:250:1250 copies of each DNA competitor, respectively) and analysis of 500 copies of HIV-1 target. Panel c and d corresponds to 2-fold competitor configuration (10:20:40:80 copies of each DNA competitor, respectively) and analysis of 10 copies of HIV-1 target. Panel e and f corresponds to a control 2-fold competitor configuration without HIV-1 target. Panel M: dye labelled ladder: 50nt, 100nt and 150 nt.

In conclusion, we have developed an integrated method for quantification of HIV-1 nucleic acids, which is suitable for automation. The methods described here can easily be extended to include HIV-1 RNA quantitation, but may also be used in gene expression analysis.

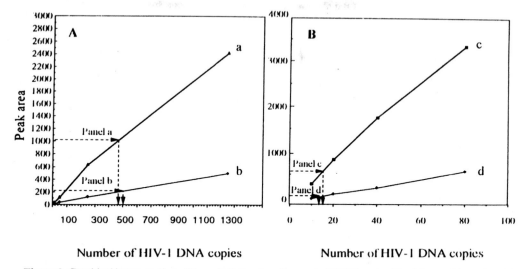

Figure 3. Graphical interpretation of figure 2. Estimates of amount of HIV-1 target, Fig. 3A: panels **a** and **b** with 5-fold competitor configuration, Fig. **3B**: panels **c** and **d** with 2-fold competitor configuration, using the calculated peak areas of the target and the competitors .

REFERENCES

1. Innis M.A. , D.H. Gelfand, J.J. Sninsky and T.J. White. PCR Protocols. Academic Press Inc., San Diego, CA (1990).
2. Brinchman J.E., J. Albert and F. Vartdal. Few infected CD4+ T cells but a high proportion of replication-competent provirus copies in asympotomatic human immunodeficiency virus type 1 infection. *J Virol.* 65: 2019-2023 (1991).
3. Becker-André M. and K. Hahlbrock. Absolute mRNA quantification using the polymerase chain reaction (PCR). A novel approach by a PCR aided transcript titration assay (PATTY). *Nucl Acids Res* 17: 9437-9446 (1989).
4. Gilliland G. , S. Perrin, K. Blanchard and H.F. Bunn. Analysis of cytokine mRNA and DNA: detection and quantification by competitive polymerase chain reaction. *Proc Natl Acad Sci USA* 87: 2725- 2729 (1990).
5. Ikonen E. , T. Manninen, L. Peltonen and A.C. Syvänen. Quantitative determination of rare mRNA species by PCR and solid-phase minisequencing. *PCR Meth Appl.* 1: 234-240 (1992).
6. Lundeberg J., J. Wahlberg and M. Uhlén. Rapid colorimetric quantification of PCR-amplified DNA. *BioTechniques* 10: 68-75 (1991).
7. Piatak M., K. Luk, B. Williams and J.D. Lifson. Quantitative competitive polymerase chain reaction for accurate quantitation of HIV DNA and RNA species. *BioTechniques* 14:70-81 (1993).

8. Clair M.H., J.L.Martin, G.Tudor-Williams, C.L.Bach, D.M.King, P.Kellam, S.D. Kemp and B.A. Larder. Resistance to ddI and sensitivity to AZT induced by a mutation in HIV-1 reverse transcriptase. *Science* 253: 1557-1559 (1991).
9. Wahlberg J., J. Albert, J. Lundeberg J., A. von Gegerfelt, K. Broliden, G. Utter G., E-M. Fenyö and M. Uhlén. Analysis of the V3 loop in neutralization-resistant human immunodeficiency virus type 1 variants by direct solid-phase DNA sequencing. *AIDS Res Hum Retrovir* 7: 983-990 (1991).
10. Vener T., M. Axelsson, J. Albert, M. Uhlén and J. Lundeberg. Quantification of HIV-1 using multiple competitors in a single-tube assay. *BioTechniques* 21: 248-253 (1996).
11. Myers G., B. Korber, J. Berkofsky, R.F. Smith and G.N. Pavlakis. Human Retroviruses and AIDS 1991. Los Alamos National Laboratory, Los Alamos, New Mexico (1991).

QUANTITATION OF *P53* TUMOR SUPPRESSOR GENE COPY NUMBER IN TUMOR DNA SAMPLES BY COMPETITIVE PCR IN AN ELISA-FORMAT

Meinhard Hahn, Volker Dörsam and Alfred Pingoud

Institut für Biochemie, FB15, Justus-Liebig-Universität Giessen, Heinrich-Buff-Ring 58, 35392 Giessen, Germany

INTRODUCTION

During the last two decades molecular tumor biology has demonstrated that most tumors are the result of multi-step mutation processes (e.g. colorectal and lung carcinoma[1,2], renal tumors[3] and tumors of the skin[4]): several mutations must accumulate in the cellular genome and modify the functions of oncogenes and/or inactivate tumor suppressor genes, which control essential cellular processes such as cell division cycle or apoptosis, before a tumor develops. The tumor suppressor gene *p53* is the gene most often affected by mutations in a large number of diverse, frequently occuring human tumors[5]. As typical for tumor suppressor genes, one copy of the *p53* gene is often inactivated in tumor cells by point mutations, while the second one is deleted. Several clinical studies showed that the inactivation of *p53* in tumors is accompanied by poor prognosis for the patients and tumor resistance against various types of chemotherapy or radiation therapy[6]. From a clinical point, therefore, a need for efficient procedures for the routine detection of *p53* mutations exists. At present, several techniques can be used to detect deletions or losses of *p53* alleles, e.g. (i) loss of heterozygosity (LOH) analysis of intragenic microsatellite polymorphisms[7,8], (ii) fluorescence *in-situ* hybridization (FISH)[6] or (iii) comparative genomic hybridization[3]. These techniques possess several disadvantages: for example in (i) the analysis of a patient is only informative in the case of a heterozygous allelotype of the polymorphism, in (ii) and (iii) one needs special, expensive equipment and great experience.

In this study we present a quantitative competitive PCR (QCPCR) system which is also able to detect *p53* deletions in tumor DNA samples but does not need laborious steps, gel electrophoresis for product separation, radioisotopes or expensive equipment. Furthermore, the QCPCR procedure presented here has the potential for automation. In this procedure a known molar amount of an internal control DNA (competitor) is added to the sample DNA. Both templates are co-amplified using the same PCR primers. The competitor DNA is of identical length as the target DNA, has identical PCR primer binding sites and more than 99 % sequence identity with the amplified genomic *p53* target sequence of the sample DNA. Nevertheless, sample and control DNA can be distinguished because of a two base pair

Modern Applications of DNA Amplification Techniques
Edited by Lassner *et al.*, Plenum Press, New York, 1997

difference which replaces the sample specific *Ssp*I restriction site (AATATT) by a control specific *Hind*III site (AAGCTT) (Figure 1). Because of the similarity in length and sequence of sample and control DNA they are amplified with identical efficiencies. This is very important, because only in the case of „ideal" competition the final ratio of sample/internal control specific PCR products will reflect the initial molar ratio of template copies, and only then is it possible to calculate the absolute *p53* copy number with high precision and reliability[8,9].

For discrimination and selective quantification of the almost identical sample/internal control specific QCPCR products we used the OLA-ELISA technique (oligonucleotide ligation assay/ enzyme linked immuno sorbent assay), a technically simple, non-radioactive, non-gel-electrophoretic assay, which we had used successfully in a previous study[10]. The product discrimination is achieved by the OLA: three oligonucleotides are hybridized to the PCR products (Figure 1). The biotinylated OLA-*up*-B is completely complementary to both PCR product species, but OLA-*down*-D and OLA-*down*-F either to sample or internal control specific PCR products. A hybridized pair of *up* and *down* oligonucleotides is ligated by T4 DNA ligase only in the case of full complementarity at the site of ligation. Thereby, the two PCR products species are distinguished and the ratio of ligation products will reflect the ratio of the competitive PCR products. The two species of ligation products are then immobilized on avidin-coated microtiter plates via their biotin-label and are quantified in digoxigenin (the hapten of the sample ligation product) and fluorescein (the hapten of the internal control ligation product), resp., specific colorimetric ELISAs. Therefore, the ratio of digoxigenin/ fluorescein signals will reflect the initial ratio of the sample/control specific *p53* DNA copies, allowing to quantify the initial *p53* copy number of the sample DNA.

MATERIALS AND METHODS

Oligodeoxynucleotides

All oligodeoxynucleotides were designed using the OLIGO™ 4.0 software (Med-Probe, Oslo, Norway) and synthesized on a MilliGen Cyclone™ Plus DNA Synthesizer (Millipore, Eschborn, Germany) using the b-cyanoethyl-phosphoramidite chemistry and reagents from Millipore. Sequences are based on the *p53* GenBank entry X54156.

Mutagenesis Primers.
QPCR-1a: 5'-CTG GCT TTG GGA CCT CTT AAC-3'
QPCR-1b: 5'-GCA GGC TAG GCT AAG CTA TGA TG-3'
QPCR-*Hind*III: 5'-TGA AGG GTG AAA GCT TCT CCA TCC AGT G-3'.

PCR Primers.
QPCR-2a: 5'-CGG CGC ACA GAG GAA GAG AAT-3'
QPCR-2b: 5'-CAA ATG CCC CAA TTG CAG GTA-3'
QPCR-3aII-FAM: 5'-*FAM* ACT AAG CGA GGT AAG CAA GC-3'
QPCR-3b: 5'-AAG AAA ACG GCA TTT TGA GT-3'.

FAM denotes the fluorophor FAM-NHS (Perkin Elmer/ABI, Weiterstadt, Germany), coupled to an Aminolink 2™ (Perkin Elmer/ABI) modified oligonucleotide. Fluorescent dye-labeling and HPLC purification are described elsewhere [7,8,11]. All primers were stored in 10mM Tris-HCl, pH 8.5 at -20°C.

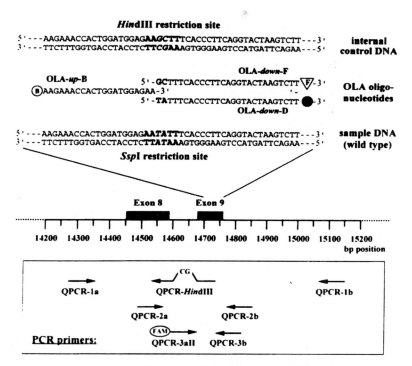

Figure 1. Scheme of the location of PCR primers (shown in the box) and of biotin or hapten labeled OLA oligonucleotides, used for PCR product discrimination. The relevant part of the human *p53* gene, including exons 8 and 9, is shown which encompasses a *SspI* restriction site (bold letters in the lower sequence), specific for sample DNA. In the internal control DNA which was generated by site directed PCR mutagenesis, the *SspI* site is replaced by a unique *HindIII* site (bold letters in the upper sequence). Sequence differences between sample and control DNA are indicated by bold italic letters. The base pair numbering is according to the GenBank entry X54156. Several oligonucleotides carry terminal modifications: the nested PCR primer QPCR-3aII is modified by the fluorophor FAM, the OLA oligonucleotides by biotin (**B**), digoxigenin (**D**) or fluorescein (**F**) label. Bold letters of OLA-down-F and OLA-down-D emphasize the 5' terminal bases, which are either complementary to the sample or to the internal control DNA.

OLA Oligodeoxynucleotides.

OLA-up-Bio:	5'-***BB*** AAG AAA CCA CTG GAT GGA GAA-3'
OLA-down-D:	5'-p-TAT TTC ACC CTT CAG GTA CTA AGT CTT ***DIG***-3'
OLA-down-F:	5'-p-GCT TTC ACC CTT CAG GTA CTA AGT CTT ***FITC***-3'.

B is a biotin-derivatized phosphoramidite (Amersham Life Science, Braunschweig, Germany). ***DIG*** (digoxigenin-3-*O*-methylcarbonyl-ε-aminocaproic acid-*N*-hydroxysuccin-imide ester;

Boehringer Mannheim) and *FITC* (fluorescein isothiocyanate; Sigma, Deisenhofen, Germany) were coupled with aminomodified oligonucleotides (3'-amino-modifier C7-CPG; MWG-Biotech, Ebersberg, Germany). After coupling, oligonucleotides were purified first by Sephadex G-25 size exclusion chromatography and later by preparative reverse phase HPLC[10,11,12]. OLA-down-D and OLA-down-F were phosphorylated using T4 polynucleotide kinase (MBI-Fermentas, St. Leon-Rot, Germany), according to the manufacturer's protocol.

Sample and Control DNA

Sample DNA. RNA-free sample DNA (wildtype sequence; $SspI^+$) was isolated either from peripheral blood lymphocytes by QIAamp Blood Kit (QIAGEN, Hilden, Germany) or from healthy or tumorous renal tissue (stored shock frozen at -70 °C) by QIAamp Tissue Kit (QIAGEN) according to the manufacturer's instructions. In all cases, DNA was eluted in 10 mM Tris-HCl, pH 8.5, and quantified by uv spectroscopy using the relation 1 $OD^{260\,nm}$ = 50 µg dsDNA/ml. For some experiments genomic blood DNA was amplified by PCR using the primer pair QPCR-1a / QPCR-1b (see Figure 1). The resulting PCR product QC-PCR-1a/ 1b ($SspI^+$) was purified employing the QIAquick PCR Purification Kit (QIAGEN) and quantified as described above. In the same way QC-PCR-2a/2b ($SspI^+$) and the fluorophor labeled DNA species QC-PCR-3aII-FAM/3b ($SspI^+$) were generated by PCR amplification, using the primer pairs QPCR-2a / QPCR-2b and QPCR-3aII-FAM / QPCR-3b, resp., purified and quantified.

Control DNA. QC-PCR-1a/1b ($HindIII^+$), the internal control DNA, was generated by site directed PCR mutagenesis using a protocol given by Perrin & Gilliland[13] in a modified version[8]. In brief, the procedure involves the following steps: first the mutagenized dsDNA fragment dsQC-PCR-1a/*Hind*III is generated by PCR amplification using the primer QPCR-1a, the mutagenesis primer QPCR-*Hind*III and genomic blood lymphocyte DNA as template. Following preparative agarose gel electrophoresis, the PCR product is isolated using the QIAEX II Gel Extraction Kit (QIAGEN). The internal control DNA is then generated using minute amounts (some ng) of genomic DNA as template, the dsDNA species dsQC-PCR-1a/ *Hind*III as a sense „megaprimer" and QPCR-1b as the antisense primer. The resulting product QC-PCR-1a/1b ($HindIII^+$) is isolated as described above and restricted by *Hind*III (Boehringer Mannheim). The two restriction fragments are isolated using the QIAEX II procedure and ligated by T4 DNA ligase (Boehringer Mannheim) according to the manufacturer's instructions. The isolated ligation product QC-PCR-1a/1b ($HindIII^+$) is reamplified by the primer pair QPCR-1a / QPCR-1b, purified using the QIAquick PCR Purification Kit and carefully quantified by uv spectroscopy. This solution of the internal control DNA and its serial dilutions in 10 mM Tris-HCl, pH 8.5 is used as the standard in quantitative competitive PCR experiments. For some studies, QC-PCR-2a/2b ($HindIII^+$) and the fluorophor labeled DNA species QC-PCR-3a/3b-FAM ($HindIII^+$) were generated by PCR amplification and isolated (see the former paragraph) using the control DNA QC-PCR-1a/1b ($HindIII^+$) as template. When preparing serial standard dilutions pipettes were gravimetrically calibrated for each pipetting step.

Polymerase Chain Reaction

PCR amplifications were performed in 0.5 ml-thin-walled tubes (Corning Costar, Boden-heim, Germany) in a thermocycler Varius V45 (Landgraf, Langenhagen, Germany). PCR standard set-ups contained 50 ng genomic template DNA (or PCR generated templates in an attomolar to nanomolar concentration range), 2.0 U *Taq* DNA polymerase (Amersham, Braun-schweig, Germany), 400 nM of each primer, 200µM of each dNTP (Pharmacia, Freiburg, Germany), 50mM KCl, 10mM Tris-HCl, pH 8.0, 1.5mM $MgCl_2$, and 0.001 % (w/v) gelatin in a final reaction volume of 50 µl and were overlaid with 70µl paraffin oil. Usually the following temperature profile was used: 35 cycles of 1min at 93 °C, 1 min at annealing temperature and 1min at 72 °C, starting the first cycle with an initial denaturation time of 5min at 93°C and finishing the last cycle with 2 min of polymerization at 72°C. Among different temperature segments there were linear ramps of 30s. The following annealing temperatures were used for the different primer pairs: QPCR-1a / QPCR-1b: 66°C; QPCR-1a / QPCR-*Hind*III: 56°C; dsQC-PCR-1a/*Hind*III / QPCR-1b: 62°C; QPCR-2a / QPCR-2b: 64°C; QPCR-3aII-FAM / QPCR-3b: 58°C.

For some experiments one-tube nested PCRs were used as described[11]. In these cases the reaction mixtures deviated from the former one only in the primer composition: they contained 12.5nM of each of the outer primers (QPCR-2a resp. QPCR-2b) and 400 nM of each of the inner primers (QPCR-3aII-FAM resp. QPCR-3b). During amplification the first 10 cycles were done at an annealing temperature of 66°C, the last 25 cycles at a temperature of 58°C. In all cases in which PCR products were analyzed by gel electrophoretical techniques the last amplification cycle was followed by 10 min of denaturation at 98°C, cooling from 98 to 70°C in 10 min, and a final quick cooling to 0°C. Thereby, it was achieved, that the double stranded PCR products show an ideal binomial distribution of the homo- and heteroduplexes of the plus and minus strands of sample and control DNA.

Analysis of PCR Products by Conventional Gel Electrophoresis and Densitometry

For discrimination of amplified sample and control DNA, 7.5µl-aliquots of a PCR product mixture were digested in a volume of 15µl either by 3U *Ssp*I or 6U *Hind*III or both enzymes, according to the manufacturer's protocol (Boehringer Mannheim). The reaction mixture was subjected to electrophoresis on 15% polyacrylamide gels. Gels were stained in an ethidium bromide solution, and photographed using an uv transilluminator (Bachhofer, Reutlingen, Germany) at 312 nm and a video camera system (INTAS, Göttingen, Germany) as described[8,11]. Product bands were quantified from the video data using the CREAM software (INTAS).

Analysis of PCR Products by Automated DNA Sequencer

Aliquots of FAM labeled PCR product mixtures (intact or cleaved with *Ssp*I and/or *Hind*III, see above) were diluted in 10mM Tris-HCl, pH8.5 (50-fold relative to the initial PCR mixture). 2.0µl of these dilutions were mixed with 0.5µl ROX-labeled internal length standard GENESCAN-2500 (Perkin Elmer/ABI) and 3µl formamide (containing 10 mg/ml blue dextran (Sigma)). The mixtures were heated at 95°C for 4min and cooled on ice very rapidly. 2µl of each sample were loaded onto a prerun 36 slot 6% polyacrylamide (acrylamide/bisacrylamide 19:1)/7M urea gel of 24cm well-to-read distance. The electrophoresis was carried out for 5.5h at a limiting voltage of 1250V, using a Model 373A DNA sequencer (Perkin Elmer/ABI). The analysis of electropherograms, the quantification of relative fragment sizes and relative molar amounts of DNA fragments were performed using GENESCAN 672 (version 1.2) collection and analysis software (Perkin Elmer/ABI) as described[7,8].

Analysis of PCR Products by OLA-ELISA

OLA. For discrimination of amplified sample and control DNA by an oligonucleotide ligation assay[8,10], 4µl of PCR product mixture were mixed with 2.8µl OLA-up-Bio, 2.8µl OLA-down-D, 2.8µl OLA-down-F (each 1µM), 70µl 2x ligation buffer (40 % (v/v) formamide, 100mM Tris-HCl, pH 7.5, 200mM NaCl, 20mM $MgCl_2$, 2mM ATP, 10 mM DTE, 10µg/ml BSA, 4mM spermidine trihydrochloride), 57.6µl H_2O and overlayed with 70µl paraffin oil. After denaturation at 95°C for 5min, the mixture was quickly cooled down to 0 °C (annealing). For ligation, 10µl T4 DNA ligase (15 Weiss U/ml) were added and the reaction mixture incubated for 30min at 37°C. The reaction was stopped by mixing with 35µl 0.25M NaOH, neutralized by 35µl 3M sodium acetate (pH 6.0), filled up to 1ml with PBST (4.3mM Na_2HPO_4, 1.4mM K_2HPO_4, 140mM NaCl, 2.7 mM KCl, 0.05% (v/v) Tween 20, pH 7.3) and analyzed immediately by an ELISA.

ELISA. Maxisorp 96 well microtiter plates (Nunc, Wiebaden, Germany) were coated with avidin and washed as described[10,11]. For each OLA experiment 80 µl of the PBST dilution were pipetted into each of six wells and incubated for 30 min under gentle shaking of the plate. After washing six times with 100µl/well PBST, the immobilized double-stranded DNA was denatured for 10 min with 100µl/well 250mM NaOH and washed again six times with PBST. To half of the wells 100µl/well of the anti-digoxigenin-Fab-POD-conjugate antibody solution were added, to the other half anti-fluorescein-Fab-POD-conjugate antibody solution (both antibodies: Boehringer Mannheim; solutions: 0.15U/ml, dissolved in PBST, containing 1% (w/v) skimmed milk powder). The plates were incubated for 30min at room temperature and washed as described above. Subsequently, the wells were filled with 80µl of a freshly prepared 3,3',5,5'-tetramethylbenzidine (Sigma) substrate solution (1 tablet dissolved in 1 ml DMSO and diluted with 9.0ml 40mM sodium acetate, 40mM sodium citrate, pH 4.4, 0.01 % (v/v) H_2O_2). After 2 - 10min, the ELISA reaction was stopped by addition of 100µl 2M H_2SO_4, such that the absorbance per well was below 1 $OD_{450 nm}$, which was measured in an ELISA reader (Dynatech, Denkendorf, Germany) using Biolinx software (Dynatech).

RESULTS/DISCUSSION

Here we present a PCR procedure for the quantification of the copy number of the human tumor suppressor gene *p53*, combined with a microtiter plate based OLA-ELISA technique for selective discrimination and quantification of competitively amplified sample and control DNA. The development of this assay comprised the (i) generation of an internal control DNA species for QCPCR, (ii) analysis and verification whether the QCPCR system shows ideal amplification properties according to theoretical considerations, (iii) optimization of the OLA-ELISA for selective PCR product quantification and finally (iv) the analysis of a tumor DNA sample which had lost one *p53* allele as determined before by an independent reference method.

GENERATION OF INTERNAL CONTROL DNA BY PCR MUTAGENESIS

It was the aim of this study to quantify the absolute *p53* template copy number in a sample DNA with high precision to be able to detect the deletion of one of the two gene copies per cell. For internal standardization we needed a control DNA fragment of identical length and almost identical sequence as the sample DNA template, in order to achieve an ideal competition during PCR. These requirements were established previously in theoretical[9] and practical[14] studies. We produced, therefore, an 871 bp DNA fragment as a control DNA by site-directed PCR mutagenesis, using a system of three PCR primers consisting of the outer primers QPCR-1a / QPCR-1a and the mutagenesis primer QPCR-*Hind*III (Figure 1) according to a modified protocol of Perrin and Gilliland[13]. The internal control fragment QC-PCR-1a/ 1b (*Hind*III[+]) includes the exons 8 and 9 of the *p53* gene. By introducing two point mutations in this fragment the single, sample specific *Ssp*I site was replaced by a new single *Hind*III site. Because the mutated DNA generated by the original protocol contained a background of *Hind*III-non-cleavable PCR products (wildtype sequence), the PCR products were cleaved by *Hind*III, the fragments isolated, ligated and re-amplified. This resulted in a homogeneous PCR product pool of *Hind*III cleavable DNA. Standard dilution series of this DNA were used as the internal control DNA in the following QCPCR titration experiments.

Analysis of the QCPCR Properties by DNA Sequencer

In the next step we had to characterize the amplification properties of the *p53* templates in the QCPCR system and to verify that the internal control DNA behaves like an ideal competitor. In this case the ratio of sample/control specific QCPCR products has to reflect precisely the initial ratio of the template DNAs, even in the late plateau phase of the PCR amplification. For this purpose we used an automated DNA sequencer operating with laser induced fluorescence. The detection of laser induced fluorescence is characterized by high sensitivity, precision and a large dynamic range. To demonstrate the feasibility of our procedure we produced in separate experiments FAM-fluorophor labeled PCR fragments in the sample and control sequence specific variant using the primer pair QPCR-3aII-FAM / QPCR-3b (Figure 1) which amplifies a sequence within the QPCR-1a / QPCR-1b defined PCR fragment. The FAM-labeled nested fragments were mixed in a defined molar ratio of sample/ control specific species and reamplified using the same primer pair (Figure 2). During the amplification process aliquots of the reaction were taken in regular intervals and cleaved by *Ssp*I and/or *Hind*III. The resulting restriction fragment mixtures were analyzed on a DNA sequencer (Figure 2a). We could show that in the logarithmic, linear and even in the very late plateau phase of the PCR, the last one defined by an amplification efficiency (for definition see Equations 1 and 2) in the range of zero (Figure 2c), the ratio of PCR fragments reflects precisely the initial molar ratio of template sequences (Figure 2b). This is the case even after 40 cycles of amplification. Hence the system fulfils the required criteria of ideal competitive amplification:

Because the former experiments were performed using sample specific PCR products as the initial template we had to show that also genomic sample DNAs show such an amplification behaviour when co-amplified with the internal control DNA QC-PCR-1a/1b (*Hind*III[+]). The amplification kinetics of sample and control specific templates can be described by Equations 1 and 2 according to Raeymaekers[9]:

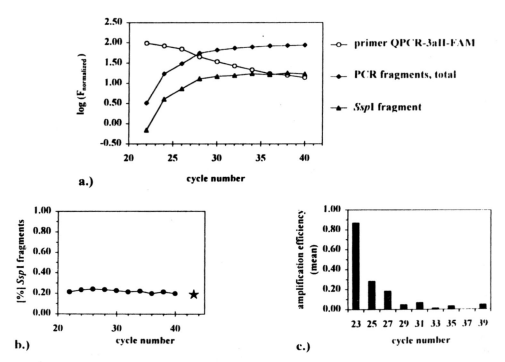

a.)

b.)

c.)

Figure 2. Analysis of the cycle dependence of PCR product accumulation (**a**), the ratio of competitively amplified PCR products of sample and control DNA (**b**) and mean amplification efficiencies (**c**). Using the *p53* specific PCR primer pair QPCR-3aII-FAM / QPCR-3b sample and control DNA specific template DNAs were amplified separately, the products isolated, quantified and mixed in an equimolar ratio (100 nM total PCR products). Aliquots of this stock solution were cleaved by *Ssp*I and/or *Hind*III. By using an automated DNA sequencer the initial molar ratio of FAM-labeled sample/control specific restriction fragments was quantified (* in panel **b**). The stock solution was diluted 10^7 fold and re-amplified using the same primer pair. In 2 cycle intervals aliquots were taken from the amplification reaction and analyzed by cleavage with restriction enzymes. The relative amounts of FAM-labeled unincorporated primer QPCR-3aII-FAM, sample specific and total PCR fragments (e.g. their relative, normalized fluorescence) were determined and plotted against the number of amplification cycles (**a**). In **b** the measured ratio of sample/control specific PCR generated DNA fragments is shown. In addition, the cycle dependent amplification efficiencies (mean values of two consecutive cycles, for definition see the text) for this experimental series are presented in **c**.

$$S_j = S_0 \times (1 + \varepsilon^S)^j \qquad (1)$$

$$C_j = C_0 \times (1 + \varepsilon^C)^j \qquad (2)$$

S:	amount of sample DNA template
C:	amount of internal control DNA template
S_0 or C_0:	initial amount of S or C template before PCR
S_i or C_i:	amount of PCR product after i PCR cycles
ε^S or ε^C:	mean amplification efficiency of S or C ($0 < \varepsilon < 1$)

In the case of an ideal competitive co-amplification of the two template species one can devide the two equations and take the logarithm (Equation 3).

$$\log(S_n/C_n) = \log S_0 - \log C_0 + n \times \log[(1 + \varepsilon^s)/1 + \varepsilon^c)] \qquad (3)$$

When both template species are amplified with identical efficiencies in each cycle, then the plot of the logarithm of the PCR product ratios against the logarithm of the initial molar amounts of added internal control DNA will result in a line with a slope of -1. Indeed, when a constant amount of RNA-free genomic lymphocyte DNA is titrated with varying amounts of competitor, coamplified and the product mixtures analyzed as described in the legend of (Figure 3), the double logarithmic plot of the data shows that the data are lying exactly on a line with a slope of -1. Our system fulfils, therefore, this requirement of an ideal quantitative competitive PCR.

Because we could determine the initial molar concentration of the *p53* copies ($S_0 = 5.91 \times 10^{-16}$ M) in a known amount of genomic DNA (50 ng per 50 µl reaction volume) it is possible to calculate the genomic length of the haploid genome (using the mean molar weight of 660 Da/base pair). We have determined this quantity to be 2.6×10^9 base pairs, a value in good agreement with data from the modern literature (for an overview see Hahn[8]). In contrast to several other QPCR techniques [15,16,17], using competitors with different size or sequence as reviewed by Raeymaekers[9], the procedure described here allows to quantify the absolute concentration of a sample DNA.

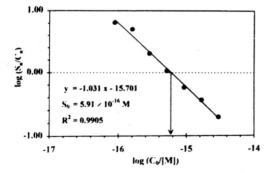

Figure 3. Double logarithmic plot of the ratio of competitively amplified PCR products (S_n/C_n) in dependence of the initial molar amount (C_0) of added internal control template. In this titration experiment 50 ng genomic lymphocyte DNA/50 µl PCR reaction volume were mixed with varying amounts (in the range of 88 aM to 3 fM) of the internal control DNA QC-PCR-1a/1b (*Hind*III⁺). After amplification for 35 cycles, using the nested PCR primer system, aliquots of the product mixtures were digested and analyzed on a DNA sequencer (see **Fig. 2**). The relative fluorescence of the different fragment species is proportional to the molar amount of fragments. Therefore, the logarithm of the ratio of sample/control specific fragment fluorescence (S_n/C_n) could be plotted against the logarithm of the initial molar amount of added internal control DNA (C_0), as described in the text. The results of the linear regression analysis of measured values and the calculated initial concentration of *p53* copies in the sample DNA (S_0) are given. The point of equimolar product ratio is marked by an arrow.

Discrimination and Quantification of PCR Products by OLA-ELISA

After having demonstrated that our amplification system is suitable for absolute quantifications of *p53* template copy numbers we had to adapt and optimize the OLA-ELISA technique for selective product discrimination and quantification. We tested different compositions of the ligation buffer (varying salt and formamide concentrations) and several concentrations of T4 DNA ligase in the OLA and different antibody dilutions in the ELISA test. The optimal protocol is described in the Materials and Methods section. It corresponds essentially to the protocol given by Friedhoff *et al.*[10]. Details of the optimization experiments are given in Hahn[8].

To demonstrate the suitability of our QPCR procedure with OLA-ELISA detection to routine analysis we analyzed DNA samples of healthy and tumorous tissue of the same renal carcinoma patient (Figure 4). For this patient we had demonstrated the deletion of one *p53* allele in all cells of the tumor sample using an independent technique (LOH analysis of several intragenic polymorphic DNA markers)[7,8]. The difference in the molar concentration of *p53* template copies per 50 ng RNA-free DNA indicates that the normal tissue contains about 2.5-fold more *p53* gene copies than the tumor tissue. Assuming that our QCPCR procedure shows a mean deviation of 10 % (results of several other titration experiments, data not shown), this means that all tumor cells have lost one *p53* gene copy and, in addition, a subset of them also the second gene copy. Because homozygous deletions cannot be shown by our reference technique (LOH detection) the QCPCR result is in agreement with the LOH results.

CONCLUSIONS

We have developed a quantitative competitive PCR procedure that couples competitive co-amplification of sample and control DNA to an ELISA based detection. The system fullfills the criteria of an ideal competitive PCR and, therefore, allows for absolute template copy quantification. Differentiation between amplified sample and control DNA is achieved by the oligonucleotide ligation assay (OLA). The procedure can be easily automated using available robotic work stations. The QPCR/OLA-ELISA technique may be adopted with appropriate modifications for other problems as quantification of virus titers or gene copy numbers in recombinant organisms.

TROUBLESHOOTING

In some cases commercially synthesized and purified digoxigenin- or fluorescein-labeled OLA oligonucleotides are not completely modified. This results in different ELISA signals for the same amount of the different OLA products and therefore in inprecise QPCR quantification results. This implies that the quality of the modified oligonucleotides has to be checked e.g. by denaturing polyacrylamide gel electrophoresis or reversed phase HPLC. In this study we have exclusively used OLA oligonucleotides which were modified and HPLC purified in our own laboratory as described [7,8,10,11,12].

In the literature there are many suitable protocols for generation by PCR of site directed mutated DNA species. It should be checked carefully that the PCR generated control DNAs are homogeneous; only then are they suitable for QPCR.

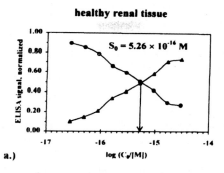
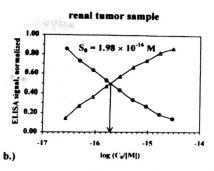

healthy renal tissue

$S_0 = 5.26 \times 10^{-16}$ M

a.) log $(C_0/[M])$

renal tumor sample

$S_0 = 1.98 \times 10^{-16}$ M

b.) log $(C_0/[M])$

Figure 4. Quantification of the *p53* copy number in a renal clear cell carcinoma by competitive QPCR and OLA/ELISA detection. RNA-free DNA of healthy renal tissue (**a**) or a renal tumor sample (**b**) of the same patient were titrated with *p53* specific internal control DNA (in the range of 27 aM to 3 fM) as described in the legend to Fig. 3. Using the primer pair QPCR-3a/QPCR-3b the template mixtures were amplified for 35 cycles. Afterwards the product species were discriminated and quantified using the OLA-ELISA technique. The normalized ELISA signals. (•) sample specific; (▲) internal control specific) are plotted against the logarithm of the initial amount of added internal control template (C_0). The point of intersection of the curves is identical with the point of equimolar initial concentrations of the two template species. In both plots this point is marked by an arrow, also shown are the initial molar concentrations of sample specific *p53* copies (S_0).

When developing a new OLA-ELISA system one has to secure a high specificity of the OLA as well as a high signal intensity of the ELISA. For a new system it might be necessary to check and to optimize the OLA buffer composition for these purposes. In the case of malfunction one should vary first the concentration of formamide and/or salt. In all cases tested this lead to suitable assay conditions. In addition one should check the ELISA conditions (e.g. dilution of antibodies, biotin binding capacity of the avidin-coated microtiter plates and blocking conditions) to achieve a robust, sensitive and precise assay.

It is an inherent unfavourable property of absorbance signal producing ELISA techniques that only in a limited concentration range (roughly one order of magnitude) a reliable linearity is given between absorbance signal and sample concentration. Therefore in future it should be a benefit to use ELISA systems which produce lumiscent or fluorescent products [18] and to use time-resolved fluorometry [19] or chemiluminescence [20], resp., for quantification of the OLA products because these techniques have a higher dynamic range of signal intensity.

When analyzing tumor samples it is an important prerequisite that the tissue sample contains a homogenous population of tumor cells, which is contaminated as little as possible by normal *p53* diploid cells. Therefore, if possible, one should analyze microdissected tumor cell samples.

ACKNOWLEDGMENTS

This work was generously supported by Boehringer-Mannheim and in part by a Ph.D. fellowship of the Deutsche Forschungsgemeinschaft, Graduiertenkolleg „Molekulare Patho-physiologie des Zellwachstums" (No. 120/6-1) to M.H.. We are grateful to Dr. Peter Friedhoff, Hamburg, and Dr. Heiner Wolfes, Hannover, for synthesis of oligonucleotides as well as Dr.

Jürgen Serth, Medizinische Hochschule Hannover for providing tumor samples.

REFERENCES

1. Yokota J., and T. Sugimura. Multiple steps in carcinogenesis involving alterations of multiple tumor suppressor genes. *FASEB J,* 7:920 (1993).

2. Fearon E.R., and B. Vogelstein, A genetic model for colorectal tumorigenesis. *Cell,* 61:759 (1990).

3. Moch H., J.C. Presti Jr., G. Sauter, N. Buchholz, P. Jordan, M.J. Mihatsch, and F.M. Waldman. Genetic aberrations detected by comparative genomic hybridization are associated with clinical outcome in renal cell carcinoma. *Cancer Res,* 56:27 (1996).

4. Ziegler A., A.S. Jonason, D.J. Leffell, J.A. Simon, H.W. Sharma, J. Kimmelman, L. Remington, T. Jacks, and D.E. Brash. Sunburn and p53 in the onset of skin cancer. *Nature,* 372:773 (1994).

5. Nigro J.M., S.J. Baker, A.C. Preisinger, J.M. Jessup, R. Hostetter, K. Cleary, S.H. Bigner, N. Davidson, S. Baylin, P. Devilee, T. Glover, F.S. Collins, A. Weston, R. Modali, C.C. Harris, and B. Vogelstein. Mutations in the *p53* gene occur in diverse human tumour types. *Nature,* 342:705 (1989).

6. Döhner H., K. Fischer, M. Bentz, K. Hansen, A. Benner, G. Cabot, D. Diehl, R. Schlenk, J. Coy, S. Stilgenbauer, M. Volkmann, P.R. Galle, A Poustka, W. Hunstein, and P. Lichter. p53 gene deletion predicts for poor survival and non-response to therapy with purine analogs in chronic B- cell leukemias. *Blood,* 85:1580 (1995).

7. Hahn M., S.E. Matzen, J. Serth, and A. Pingoud. Semiautomated quantitative detection of loss of heterozygosity in the tumor suppressor gene *p53. BioTechniques,* 18:1040 (1995).

8. Hahn M. „Entwicklung neuer PCR-Analyseverfahren für das humane Tumorsuppressorgen *p53,*" Ph.D. theses, Justus-Liebig-Universität, Giessen (1996).

9. Raeymaekers L. Quantitative PCR: theoretical considerations with practical implications. *Anal Biochem,* 229:582 (1993).

10. Friedhoff P., M. Hahn, H. Wolfes, and A. Pingoud. Quantitative polymerase chain reaction with oligodeoxynucleotide ligation assay/enzyme-linked immunosorbent assay detection. *Anal Biochem,* 215:9 (1993).

11. Hahn M., V. Dörsam, P. Friedhoff, A. Fritz, and A. Pingoud. Quantitative polymerase chain reaction with enzyme-linked immunosorbent assay detection of selectively digested amplified sample and control DNA. *Anal Biochem,* 229:236 (1995).

12. Landgraf A,, B. Reckmann, and A. Pingoud. Direct analysis of polymerase chain reaction products using enzyme-linked immunosorbent assay techniques. *Anal. Biochem,* 198:86 (1991).

13. Perrin S., and G. Gilliland. Site-directed mutagenesis using asymmetric polymerase chain reaction and a single mutant primer. *Nucleic Acids Res,* 18:7433 (1990).

14. McCulloch R.K., C.S. Choong, and D.M. Hurley. An evaluation of competitor type and size for use in the determination of mRNA by competitive PCR. *PCR Methods Applic,* 4:219 (1995).

15. Becker-André M., and K. Hahlbrock. Absolute mRNA quantification using the polymerase chain reaction (PCR). A novel approach by a PCR aided transcript titration assay (PATTY). *Nucleic Acids Res,* 17: 9437 (1989).

16. Gilliland G., S. Perrin, K. Blanchard, and H.F. Bunn. Analysis of cytokine mRNA and DNA:

detection and quantitation by competitive polymerase chain reaction. *Proc Natl Acad Sci USA*, 87:2725 (1990).

17. Siebert P.D., and J.W. Larrick. Competitive PCR. *Nature* 359:557 (1992).

18. Bortolin S., T.K. Christopoulos, and M. Verhaegen. Quantitative polymerase chain reaction using a recombinant DNA internal standard and time-resolved fluorometry. *Anal Biochem*, 68:834 (1996).

19. Samiotaki M., M. Kwiatkowski, J. Parik, and U. Landegren. Dual-color detection of DNA sequence variants by ligase-mediated analysis. *Genomics*, 20:238 (1994).

20. DiCesare J., B. Grossman, E. Katz, E. Picozza, R. Ragusa, and T. Woudenberg. A high-sensitivity electrochemiluminescence-based detection system for automated PCR product quantitation. *BioTechniques*, 15:152 (1993).

STANDARDISATION OF MESSENGER RNA QUANTIFICATION USING AN RT-PCR METHOD INVOLVING CO-AMPLIFICATION WITH A MULTI-SPECIFIC INTERNAL CONTROL

David Shire, Pascale Legoux and Adrian J. Minty

Sanofi Recherche, Centre de Labège, 31676 Labège, France

INTRODUCTION

The quantification of the amount of a specific mRNA in a given tissue or cell culture is not a trivial task. Generally, the quantities of total RNA available are too low for direct quantification and amplification-based methods have to be used. Reverse transcription coupled with the polymerisation chain reaction (RT-PCR) is a popular method and can give acceptable results provided known quantities of an internal control are present throughout the procedure to compensate for variations in reverse transcriptase and DNA polymerase efficiency.

The internal controls that have been used up to now can be divided into two categories. The first consists of controls that are directly derived from the cellular template by deleting or adding some internal nucleotides and using the small size or sequence difference to distinguish the two amplicons after amplification. The advantages of using such a control are the ease of obtention of the control and the identical amplification efficiencies for the cellular template and the control. The disadvantages are the formation of heteroduplexes between the target and control amplicons; potential crossing-over during amplification and the restriction that only one target can be measured per control. The last disadvantage is particularly acute if one wishes to study complex systems such as the cytokine network in which a large number of different transcripts can be induced upon cell activation and where the nature of the transcripts and their concentration depends upon the cell type and the activation employed. A direct comparison of the levels of these transcripts necessitates a more global approach than the one afforded by the one control/one target method.

To meet this need Wang et al[1] introduced the concept of synthetic multi-specific internal controls consisting of a series of upstream priming sites for cytokine and growth factor cDNAs organised head-to-tail, followed by a linker region leading to a series of downstream priming sites for the same targets placed in the same order. The constructs are inserted into a plasmid (Figure 1A) behind a promoter allowing a synthetic control RNA (st-RNA) to be produced that can be used to spike total cellular RNA (cell-RNA). Reverse transcription and

Modern Applications of DNA Amplification Techniques
Edited by Lassner *et al.*, Plenum Press, New York, 1997

PCR conditions are subsequently identical for both species. The advantages of the multi-specific approach are that several targets are measurable simultaneously without the problems of heteroduplex formation or cross-over. The two principal problems are potential differences in RT and PCR efficiency between the control and target sequences. These problems will be addressed here.

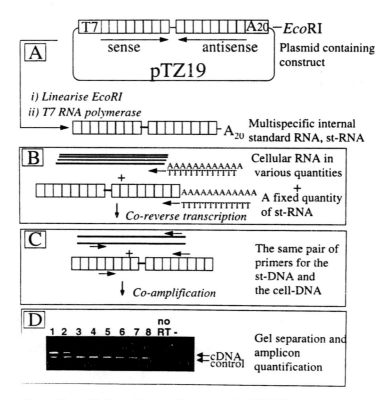

Figure 1. The use of a multi-specific internal control for quantitative RT-PCR

MATERIALS AND METHODS

Linearisation of the plasmid

Plasmid pQB-3 (4µg) constructed as described[2] was dissolved in 30µl of *Eco*RI buffer (50 mM NaCl, 0.1M Tris-HCl, pH 7.9; 5mM MgCl$_2$, 0.0025% Triton X100) and incubated with *Eco*RI (40 U) for 1 h at 37°C. A further 10 µl of the above buffer and 2µl of *Eco*RI was added and the mixture was left for a further 1 h at 37°C. It was extracted with 1 volume phenol saturated with 10mM Tris pH 8, 1mM EDTA, then with 1 volume chloroform and precipitated with 2 volume ethanol at -20°C and the pellet dried. It was dissolved in 40µl water and 1µl was electrophoresed on a 1% agarose gel, with 100ng of the original plasmid to ensure complete linearisation and to assess the quantity of recovered plasmid. The linearised plasmid was stored lyophilised or in solution at -20°C.

Transcription and cRNA purification

5x T7/T3 reaction buffer (4μl, Epicentre Technologies), 25 mM ATP, CTP, GTP, UTP (6μl of NTP mixture, final concentration of each 7.5mM), 100 mM dithiothreitol (2μl, 10mM final) were mixed and linearised plasmid (1μg in 8μl water) and Ampliscribe T7 RNA polymerase (2μl, Epicentre Technologies (by far the most efficient RNA polymerase we have tried) were added. After incubation for 2h at 37°C RNase-free DNase (1μl) was added , incubation continued for 15min at 37°C. The solution was cooled in ice and extracted with phenol-chloroform as above.

Biogel P10 (2g) was swollen in 100 ml TE buffer (10mM Tris pH 8, 1mM EDTA), the supernatant was removed and 100 ml TE was added. The suspension of 2/3 gel, 1/3 buffer in two 50 ml Falcon tubes was stocked at 4°C. In a 50 ml tube borosilicate glass beads (100-200 μm diameter, 10ml) were washed successively with 1M HCl, 1M NaOH, 1M Tris pH 8, decanting floating glass debris at each washing. The beads were treated with diethyl-pyrocarbonate (10μl/100 ml water, Sigma) overnight at room temperature before autoclaving for 20 min at 120°C. The bottom of a microtube was pierced with a hot sterile needle. With the aid of a truncated sterile plastic cone 10μl of the glass beads were placed in the tube, with a minimum of liquid. The tube was completely filled with P10, the microtube placed on a truncated Eppendorf tube inserted into a conventional centrifuge tube and centrifuged for 5 min at 5,000 rpm on a swinging rotor. It was ensured that the resin remained moist. The microtube was placed on a sterile 1.5 ml Eppendorf tube posed on the centrifuge tube. 20-30μl of cRNA solution was pipetted onto the resin and the tube was centrifuged for 5min at 5,000rpm on a swinging rotor. The eluant was lyophilised.

cRNA quantification and quality control

The pellet was dissolved in 100μl bidistilled water. 2μl was pipetted into 98μl water and a UV- spectrum in a 50μl quartz cuvette was obtained. The 260 nm/280 nm ratio was confirmed to be 1.6-1.8.

The quantity of cRNA $= OD_{260nm}/ 1000 \times 50/2 \times 40$ μg/μl (assuming 1 OD_{260nm}/ml = 40μg)

$$= OD_{260nm} \text{ μg/μl}$$

In several preparations from different plasmids the yields obtained were between 50 and 90 μg of pure cRNA from 1 μg of plasmid.

A 20x RNA buffer containing 40ml 1M MOPS, pH 7.4, 4ml 0.5 M EDTA, 156ml water was prepared, filtered and stocked at -20°C. cRNA (1-2 μg) was diluted in 1x RNA buffer (5μl) and 1.33x blue dye (15μl of a solution containing formamide (674μl), formaldehyde (216μl), 1 M MOPS pH 7.4 (30μl), 1 M EDTA (3μl), 10% SDS (10μl), glycerol (57μl) and 1% bromophenol blue (10 μl))was added. The mixture was denatured for 5min at 65°C and cooled quickly in ice. The totality of the cRNA and a size ladder were separated on a 1% agarose gel for 30min at 50 V. After staining the gel for 5min in ethidium bromide solution a second short electrophoresis eliminated excess dye, resulting in a better visualisation of the single band of pure cRNA that was obtained.

The pure cRNA purified was stored at -20°C. It was used at 20 fg/PCR in all experiments. A stock solution of 200fg/μl in bidistilled water (1μl for 10 simultaneous PCRs) has been used for several years without cRNA deterioration despite repeated freezing and defreezing, care being taken however to avoid contamination.

Peripheral blood mononuclear cell RNA preparation

Non-adherent cells were prepared and stimulated as described[3] for 5h with phorbol 12-myristate 13-acetate (PMA, 10ng/ml, Sigma), phytohaemagglutinin (PHA, 5μg/ml, Sigma), anti CD28 (1μg/ml) and cyclosporin A (1μg/ml). RNA was extracted using the method of Chomczynski and Sacchi[4].

RT-PCR quantification

Before undertaking the quantification as such, it is highly advisable to ascertain the presence of the targeted RNA in the mixture to be analysed, together with its approximate quantity. The objective is to obtain an easily visible ethidium bromide-stained band on a gel after 30 amplification cycles. We generally fix the quantity of total RNA for this preliminary test at 50ng (or about 0.5ng polyA[+] RNA) per target. This allows one to distinguish between poorly- and well-expressed RNAs. If the resulting band is very intense or weak, further preliminary tests with 5ng or 500ng, respectively, are advisable. If more than 500ng appears to be necessary, then it must be concluded that the RNA is too rare to be quantifiable and the results will be meaningless. For cytokine messengers in stimulated cells, the preliminary test can be carried out on 5ng of total RNA. In each case the total RNA sample may be spiked with 20fg cRNA. The usual control reaction of PCR without RT can be carried out here to verify the absence of contaminating genomic DNA in the RNA preparations. Contamination may also be evident during amplicon separations, since most of the primers surround splice sites in the cellular targets, but some of the exons are very large.

PRELIMINARY TEST

Reverse Transcription

An aqueous solution (14 μl) of total RNA (50ng), and dT_{12-18} (100ng) was heated for 15 min at 65°C. After cooling to RT (~15 min), reverse transcriptase buffer (4μl, BRL)), 0.1M DTT (2μl), 20 mM of each dNTP (0.5μl of DNA polymerisation mixture, LKB-Pharmacia), RNase inhibitor (0.5μl, 50U/μl, Promega) and Superscript reverse transcriptase (1μl, 200 U/μl, BRL) were added and the solution heated for 1h at 37°C. The reaction was stopped at 95°C for 5 min, then ice cooled.

PCR

Primers (100ng of each), PCR buffer (5μl, Perkin-Elmer), dNTP mix (0.8μl of a mixture containing 5mM of each), AmpliTaq (0.2μl, 5U/μl, Perkin-Elmer) and water to 50μl were added. Amplification was for 30 cycles with for 1min at 94°C, 1min at 55°C and 1min at 72°C.

The solution was concentrated and separated on a 2% agarose gel at 100 V in 1x Tris-borate. The intensity of the band gave an idea of the quantity of total RNA necessary for the quantification experiments.

The total RNA was tested for the presence of many mRNAs by multiplying the initial quantities by the number of mRNAs to be assessed and dividing up the resulting cDNA mixture before adding the appropriate primers. Grouping together mRNAs of similar abundance can be time-saving during quantification experiments.

Quantification

The appropriate quantity of total RNA (multiplied by the number of PCRs to be carried out) containing 20fg cRNA (multiplied by the number of PCRs to be carried out) was reverse transcribed as above. The mixture of control cDNA (st-cDNA) and cellular cDNA (cell-cDNA) was divided into x fractions in PCR tubes. The PCR reactions were carried out as above in the presence of [^{32}P]dCTP (1μCi/PCR, Amersham, 3,000 Ci/mmol, 10μCi/μl). Half the PCR mixture containing 10% blue dye solution was separated on a 10% polyacrylamide gel (Miniprotean II Ready gel, Biorad) at 60V for about 3h. The gel was exposed in a PhosphorImager cassette for at least 2h or the amplicons were excised and and measured by Cerenkov counting.

For each lane, the ratio R was calculated, where

$$R = \frac{\text{no. of pixels (or cpm) of the cellular amplicon}}{\text{no. of pixels (or cpm) of the control amplicon}} \times F*$$

$$\text{and} \quad F = \frac{\text{amplicon length for cell-cDNA}}{\text{no. of dC + dG in cell-cDNA}}$$

If R<1 the RT-PCR was repeated using a larger quantity of total RNA; if R>1 the total RNA was decreased. The objective was to obtain R as near 1 as possible. To obtain the quantity of total RNA that contains 20fg of the mRNA targeted, a graph was drawn of R against total RNA (ng) on a log-log scale as shown in figure 2.

Separation of amplicons can be accomplished on agarose. This is quite acceptable once it is certain that both st-cDNA and cell-cDNA amplify with equal efficiency. A plot of slope 1, obtained on polyacrylamide gel separation, will not be observed, as we have previously demonstrated[2]. However, the point of equivalence, where st-amplicon = cell-amplicon will be identical[2].

RESULTS AND DISCUSSION

Many multi-specific internal controls have been constructed by us[2,5,6] and by others[1,7-13] and all follow the similar basic design shown in figure 1A. The plasmids containing the mRNA targets shown in Table 1 are available from the authors and from Dr. Gutierrez-Ramos[10]. In the authors' constructs the primers are highly standardised in that nearly all are 20 nucleotides in length and designed with a 50% dC/dG content in order to have similar Tm values, thereby allowing one to use identical PCR conditions for each target. Wherever possible, the primers surround splice sites to avoid amplifying contaminating genomic DNA and to help its detection. Each pair of primers is separated by the same distance in each construction e.g. 410 bp in pQB-3. Exactly the same priming sites are found in the respective mRNAs, but the distance between them differs from that in the standard. This size difference facilitates separation of the amplicons after PCR. The constructs are contained in a pTZ plasmid, between a T7 promoter and a polydA stretch ending in a unique *Eco*RI site. The constructs contain other unique restriction sites, to allow modifications by cassette mutagenesis[14] or by PCR[15]. The 21 bp linker regions in all of the constructs are identical and a 'universal' probe can thus be used to identify standard amplicons. The *Eco*RI site is used for linearisation of the plasmids before transcription by T7 RNA polymerase to produce st-RNA.

Plasmid	Species	Target sequences
pQA-1	Human	IL-1b, IL-2, IL-3, IL-4, IL-5, IL-6, IL-8, TNFα, IFN-γ, GM-CSF, G-CSF, CSF1, IL-2rec, β2-microglobulin
pQB-1	Human	c-fos, Krox-20, Krox-24, jun B, IL-8, MCP-3, MCP-1, Gro-g, MIP1a, MIP1b, c-jun, jun D, β2-microglobulin, β-actin
pQB-3	Human	c-fos, Krox-20, Krox-24, jun B, IL-8, MCP-3, MCP-1, Gro-g, MIP1a, MIP1b, IL-10, IL-4, IL-12p40, IL-13 , β2-microglobulin, β-actin
pMus-3	Mouse	IL-1a, IL-1b, IL-2, IL-3, IL-4, IL-5, IL-6, TNFα, IFN-γ, GM-CSF, IL-2p35, IL-2p40, IFN-γ rec, IL-10, IL-13, β2-microglobulin
pRat6	Rat	IFN-γ, IL-10, IL-1ra, IL-1β, IL-6, , MBP, BDNF, NT-3, VGF, NT-4/5, IL-2, CSF-1, IL-1α, MCP-1, TGFmp, NGFβ, TNFα, TGFβ1, β2-microglobulin, β-actin
pSPC110	Mouse (receptors)	IL-1R-1, IL-1R-2, IL-2Ra, IL-2Rβ, IL-2Rγ, IL-3R, AIC2A, AIC2B, IL-4R, IL-5Rα, IL-6R, gp130, IL-7R, IL-9R, GM-CSF R, G-CSF R, EPO R, c-KIT, TNFR1, TNFR2, INFGR, MCSFR

Table 1. Multi-specific internal control plasmids and their target sequences

The st-RNA can be used for three purposes, i) by spiking an RNA sample one can simply verify in a qualitative manner that the RT-PCR procedure is working correctly; ii) for preliminary screening of a large number of RNA samples one can compare the relative quantities of selected targets ; a fixed quantity of total RNA from each sample is spiked with a fixed quantity of st-RNA and the intensities of the resulting amplicons are compared in a semi-quantitative manner. For many purposes, this procedure is sufficient in itself. iii) For precise quantitation, some careful controls are necessary, as described below. In this presentation we will only consider the third use of the standards, which in any case encompasses the first two.

The principle of the method is shown in figure 1. A known quantity of st-RNA is added to a known quantity of total cell-RNA (Figure 1B). Co-reverse transcription gives a mixture of st-cDNA and cell-DNA. The primer pair is added and the cDNA mixture is coamplified using the same primers (Figure 1C). The two amplicons (st-amplicon and cell-amplicon) are separated electrophoretically (Figure 1D) and the bands are quantitated. By separating the cDNA mixture into several aliquots different primer pairs can be used in parallel PCRs and separated on the same gel as previously shown[16]. Alternatively, cell-RNA can be varied, keeping st-RNA constant and one or more mRNAs can be quantitated simultaneously. One important advantage of this strategy is that after mixing st-RNA and cell-RNA the whole RT-PCR procedure is carried out in common for the two RNAs and the final amplicon ratio is directly proportional to the original st-RNA/cell-RNA ratio.

The objective of the method is to find the point of equivalence, where the two amplicon bands are of equal intensity, showing that the st-RNA added at the outset matches the quantity of the mRNA targeted. In order to achieve this, several rounds of RT-PCR are necessary, keeping st-RNA constant and using different quantities of cell-RNA. Finally, a log-log plot of [cell-amplicon]/[st-amplicon] against [total cell-RNA] should give a straight line of slope 1 as we showed previously[2,16]. The amount of cell-RNA containing the quantity of mRNA equivalent to st-RNA is obtained from the graph from the point where [cell-amplicon]/[st-amplicon] = 1 on the log-log scale. Raemaekers[17] has presented an excellent theoretical treatment of competitive PCR.

To illustrate the method, figure 2 shows the results obtained for the measurement of IL-13 mRNA in peripheral blood mononuclear cells following several different treatments. Unstimulated cells contain very low amounts of the mRNA, as revealed by the low ratio of cell-amplicon/st-amplicon, even starting with 300ng of total cellular RNA. From the point of equivalence it can be estimated that there is roughly 20fg of IL-13 mRNA in 500ng of total cellular RNA, a negligable quantity. However, stimulation of cytokine transcription by PMA or double stimulation with PMA and anti-CD28 or PMA and PHA-P reveals the presence of 20fg of the mRNA in 10ng, 1ng and 2ng total RNA, respectively. The PMA/PHA-P stimulation is reduced 3-fold in the presence of the immunosuppressor CsA. It should be noted that this type of experiment is the most simple to interpret, since it relates to cells in culture under easily controlled conditions. Tissue studies are much more difficult to undertake, particularly as regards the quantity of total RNA extracted and the final results should be adjusted by referring to the amount of a housekeeping gene present in the extract. The choice of such a gene is difficult and depends highly on the tissues being examined.

TROUBLESHOOTING

The two major problems of quantitative RT-PCR concern the efficacities of the reverse transcription and the amplification. There are three possibilities for the choice of primer for the reverse transcription, dT_{12-18}, random priming and specific priming with the antisense primer. We have adopted dT_{12-18} priming. The use of this primer assumes the presence of a polyA stretch in the mRNA to be assayed (it is present in the internal standard cRNA), which is generally the case in functional mRNAs. An objection may be raised that mRNAs may be incompletely reverse-transcribed because of long 3'-UTR or due to pauses caused by secondary structures. The distances between the polyA region and the sense primers we have chosen range from 400 to 1600nt, with the great majority being less than 900nt. Outstanding exceptions are c-jun (2100nt), Krox20 (2600nt) and CSF-1 (3600nt). The maximum distance in the controls is 600 nt. As to the second point, our experience with preparing cDNA libraries has shown us that Superscript reverse transcriptase is highly efficient at producing full-length transcripts from mRNA up to at least 7kb in length.

If random priming with hexamers is chosen for the reverse transcription, it must be remembered that the whole RNA population will be transcribed, not just the 1-2% polyA$^+$ RNA, possibly resulting in increased non-specific amplification. However, satisfactory results can apparently be obtained if hexamer concentrations are kept low (0.4 ng/μg cellular RNA). A combination of dT_{12-18} and random priming has also been used. We have no personal experience in the use of specific antisense PCR primers for the reverse transcription.

It can be seen from figure 2 that the PCR plots are of unit slope, which indicates that both the control and the cellular target are amplified with identical efficiency throughout the relative concentration range employed. This provides strong evidence that the internal nucleotide sequences between the primers, although differing in both length and constitution, has no effect on amplification efficiency for this particular target. The difference in internal sequence provides the most serious concern regarding the multi-specific approach and each template has to be carefully examined. Significantly, in every case so far studied the internal sequences have had no effect on amplification efficiency (human mRNAs: IL-4, IFNγ, TNFα, IL-2rec, β2microglobulin[2], IL-5[18], IL-13, c-fos, jun B, Krox-20, Krox-24 (P. Legoux, unpublished results), rat mRNAs[6] and mouse mRNAs (F. J. Pitossi, unpublished results)).

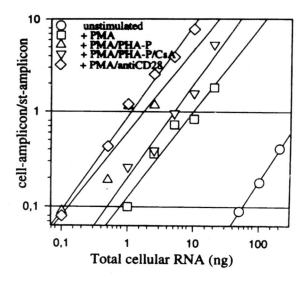

Figure 2. Log-log plot of cell-amplicon/ st-amplicon against cell-RNA for IL-13 mRNA in peripheral blood mononuclear cells

This has been established not only by quantitative plots, such as those in figure 2, but also by measuring the amount of amplican produced over a series of PCR cycles at a fixed st-DNA/ cell-DNA ratio[2]. The resulting parallelism obtained, even into the plateau phase of amplification[2,6,15,18-20], clearly shows that that the amplification efficiencies are identical. This has only to be established once for a given template, meaning that all researchers who work with the plasmids that are in free circulation can be assured of obtaining reliable results. Despite parallelism into the plateau phase, for accurate quantification we restrict measurements to the exponential phase because of the appearance of contaminating bands at high cycle numbers.

If the procedure described here is followed carefully problems should not be encountered, other than those classically encountered in RT-PCR, such as primer quality or contaminations. The main problem is that the method is labour intensive. Many strategies are presently being explored to reduce the work load and to avoide the use of radioelements[21]. Gel-based techniques include Southern analysis[22] and measurement of fluorescence intensity after ethidium bromide staining[23]. Capture strategies have been developed using paramagnetic beads and chemiluminescent quantification[18,20], and using plate assays with sandwich-based amplicon quantification[24,25]. Capillary electrophoresis[26] and HPLC[27] are also very promising techniques. A completely automated procedure is within sight.

REFERENCES

1. Wang A.M., M.V. Doyle, and D.F. Mark. Quantitation of mRNA by the polymerase chain reaction. *Proc Natl Acad Sci U S A*, 86:9717 (1989).
2. Bouaboula M., P. Legoux, B. Pésségué, B. Delpech, X. Dumont, M. Piechaczyk, P. Casellas, and D. Shire. Standardization of mRNA titration using a polymerase chain reaction method involving co-amplification with a multispecific internal control. *J Biol Chem* 267: 21830 (1992).
3. Minty A., P. Chalon, J.-M. Derocq, X. Dumont, J.-C. Guillemot, M. Kaghad, C. Labit, P. Leplatois, P. Liauzun, B. Miloux, C. Minty, P. Casellas, G. Loison, J. Lupker, D. Shire, P. Ferrara, and D. Caput. Interleukin-13 is a new human lymphokine regulating inflammatory and immune responses. *Nature,* 362: 248 (1993).
4. Chomczynski P., and N. Sacchi. Single-step method of RNA isolation by acid guanidium thiocyanate-phenol-chloroform extraction. *Anal Biochem,* 162:156 (1987).
5. Shire D., and Editorial Staff of E.C.N. An invitation to an open exchange of reagents and information useful for the measurement of cytokine mRNA levels by PCR. *Eur Cytokine Netw,* 4:161 (1993).
6. Pousset F., J. Fournier, P. Legoux, P. Keane, D. Shire, and B. Bloch. The effect of serotonin on the expression of cytokines by rat hippocampal astrocytes. *Mol Brain Res,* 38: 54 (1996).
7. Platzer C., G. Richter, K. Überla, W. Müller, H. Blöcker, T. Diamantstein, and T. Blankenstein. Analysis of cytokine mRNA levels in interleukin-4-transgenic mice by quantitative polymerase chain reaction. *Eur J Immunol,* 22: 1179 (1992).
8. Feldman A.M., P.E. Ray, C.M. Silan, J.A. Mercer, W. Minobe, and M.R. Bristow. Selective gene expression in failing human heart. Quantification of steady-state levels of messenger RNA in endomyocardial biopsies using the polymerase chain reaction. *Circulation,* 83: 1866 (1991).
9. Platzer C., S. Ode-Hakim, P. Reinke, W.-D. Döcke, R. Ewert, and H.-D. Volk. Quantitative PCR

analysis of cytokine transcription patterns in peripheral mononuclear cells after anti-CD3 rejection therapy using two novel multispecific competitor fragments. *Transplantation*, 58: 264 (1994).

10. Jia G.Q., and J.C. Gutierrez-Ramos. Quantitative measurement of mouse cytokine mRNA by polymerase chain reaction. *Eur Cytokine Netw* 6: 253 (1995).

11. Siegling A., M. Lehmann, C. Platzer, F. Emmrich, and H.-D. Volk. A novel multispecific competitor fragment for quantitative PCR analysis of cytokine gene expression in rats. *J Immunol Methods*, 177: 23 (1994).

12. Farrar J.D., and N.E. Street. A synthetic standard DNA construct for use in quantification of murine cytokine mRNA molecules. *Mol Immunol*, 32: 991 (1995).

13. Tarnuzzer R.W., S.P. Macauley, W.G. Farmerie, S. Caballero, M.R. Ghassemifar, J.T. Anderson, C.P. Robinson, M.B. Graut, M.G. Humphreys-Beher, L. Franzen, A.B. Peck, and G.S. Schultz. Competitive RNA templates for detection and quantitation of growth factors, cytokines, extracellular matrix components and matrix metalloproteinases by RT-PCR. *Biotechniques*, 20: 670 (1996).

14. Shakhov A.N. New derivative of pmus for quantitation of mouse IL-12 (p35, p40), il-10 and IFNγ-R mRNAs. *Eur Cytokine Netw*, 5: 337 (1994).

15. Pitossi F.J., and H.O. Besedovsky. A multispecific internal control (pRat6) for the analysis of rat cytokine mRNA levels by quantitative RT-PCR. *Eur Cytokine Netw*, 7: 377 (1996).

16. P. Legoux P., C. Minty, B. Delpech, A.J. Minty, and D. Shire. Simultaneous quantitation of cytokine mRNAs in interleukin-1b stimulated U373 human astrocytoma cells by a polymerase chain reaction method involving co-amplification with an internal multi-specific control. *Eur Cytokine Netw* 3: 553 (1992).

17. Raeymaekers L. Quantitative PCR: Theoretical considerations with practical implications. *Anal Biochem*, 214: 582 (1993).

18. Vandevyver C., and J. Raus. Quantitative analysis of lymphokine mRNA expression by an automated, non-radioactive method. *Cell Mol Biol*, 41: 683 (1995).

19. Siebert P.D., and J.W. Larrick. PCR MIMICS: competitive DNA fragments for use as internal standards in quantitative PCR. *Biotechniques*, 14: 244 (1993).

20. Motmans K., J. Raus, and C. Vandevyver. Quantification of cytokine messenger RNA in transfected human T cells by RT-PCR and an automated electrochemiluminescence-based post-PCR detection system. *J Immunol Methods*, 190: 107 (1996).

21. Jenkins F.J. Basic methods for the detection of PCR products. *PCR Methods Appl*, 3: S77 (1994).

22. Bickel M., S.M. Nöthen, K. Freiburghaus, and D. Shire. Chemokine expression in human oral keratinocyte cell lines and keratinized mucosa. *J Dental Res*, in press (1996).

23. Kopf M., F. Brombacher, G. Kienzle, K. Lefranc, C. Humborg, and W. Solbach. IL-4-deficient mice are resistant to infection with *Leishmania major*. *J Exp Med*, in press (1996).

24. Hockett R.D., Jr., K.M. Janowski, and R.P. Bucy. Simultaneous quantitation of multiple cytokine mRNAs by RT-PCR utilizing plate based EIA methodology. *J Immunol Methods*, 187: 273 (1995).

25. Zou W.P., I. Durand-Gasselin, A. Dulioust, M.C. Maillot, P. Galanaud, and D. Emilie. Quantification of cytokine gene expression by competitive PCR using a colorimetric assay. *Eur Cytokine Netw*, 6: 257 (1995).

26. Williams S.J., C. Schwer, A.S.M. Krishnarao, C. Heid, B.L. Karger, and P.M. Williams. Quantitative competitive polymerase chain reaction: analysis of amplified products of the HIV-1 *gag* gene bycapillary electrophoresis with laser-induced fluorescence detection. *Anal Biochem*, 236: 146 (1996).

27. Haywardlester A., P.J. Oefner, and P.A. Doris. Rapid quantification of gene expression by competitive RT-PCR and ion-pair reversed-phase HPLC. *Biotechniques,* 20: 250 (1996).

QUANTITATIVE ANALYSIS OF HUMAN DNA SEQUENCES BY SOLID-PHASE MINISEQUENCING

Ann-Christine Syvänen

Department of Human Molecular Genetics, National Public Health Insitute, Mannerheimintie 166, 00300 Helsinki, Finland

INTRODUCTION

The PCR technique provides highly specific and sensitive means for analysing nucleic acids. A drawback of the PCR based methods is, however, that they do not allow direct quantification of a nucleic acid sequence. This problem originates from the fact that the efficiency of PCR depends on the amount of template sequence present in the sample, and therefore the amplification is exponential only at low template concentrations[1.] Due to this "plateau effect" of the PCR, the amount of the amplification product does not reflect directly the original amount of the template. Moreover, subtle differences in the reaction conditions, such as material from biological samples, may cause significant sample to sample variation in the final yield of PCR product.

The problem of performing accurate quantitative PCR analyses has been addressed by two principal approaches. A quantitative PCR result can be obtained by "kinetic" PCR, in which the amplification process is monitored at numerous time or concentration points[2-3]. This approach is laborious, and requires a sensitive method for detecting the PCR products at a stage during which the amplification still proceeds exponentially. The other approach, denoted "competitive" PCR makes use of an internal quantification standard that is co-amplified in the same reaction as the target sequence[4-6]. Because the efficiency of the amplification is affected by the sequence of the PCR primers, as well as the size and, to some extent, by the sequence of the template, the internal standard used should be as similar to the target sequence as possible to ensure that the ratio between the two sequences remains constant throughout the amplification. An ideal PCR quantification standard differs from the target sequence only at one nucleotide position, by which the two sequences can be identified and quantified after the amplification.

The determination of the relative amounts of the PCR products originating from the target and standard sequence allows calculation of the initial amount of the target sequence. In situations where two target sequences are present as a mixture in a sample it is easy and often sufficient to measure the relative amounts of them. To be able to determine the absolute amount of target sequence, it is necessary to add a known amount of standard sequence to the

Modern Applications of DNA Amplification Techniques
Edited by Lassner *et al.*, Plenum Press, New York, 1997

sample before the amplification. In this case a measure of the amount of the analysed sample, such as the number of cells or the total amount of DNA, RNA or protein, is needed.

We have developed a solid-phase minisequencing method for the identification of point mutations or nucleotide variations in human genes[7]. The method distinguishes between two sequences that differ from each other only at a single nucleotide. Therefore it is an ideal tool for quantitative analysis of DNA[8] and RNA[9] sequences by competitive PCR. This communication describes the principle of the solid-phase minisequencing method and applications of it for quantitative analysis of human DNA sequences. Examples of determining the relative amounts of two DNA sequences differing from each other at a single nucleotide that are present as a mixture in a sample, as well as the determination of the absolute amount of a DNA sequence with the aid of an internal standard will be given.

MATERIALS AND METHODS

Two PCR primers and one detection step primer for the minisequencing reaction are required. One of the PCR primers is biotinylated in its 5'-end during its synthesis. The PCR primers should amplify a fragment, preferably between 50 and 500 base pairs in size, containing the variable nucleotide position(s). The PCR primers should be 20-23 nucleotides long, have similar melting temperatures and non-complementary 3'-ends[10]. The minisequencing detection step primer is 20 nucleotides long and complementary to the biotinylated strand of the PCR product immediately 3' of the variable nucleotide position (Figure 1).

Various types of DNA samples, treated as is suitable for PCR amplification can be analysed[11]. PCR is carried out according to a standard protocol[12], except that the concentration of the biotinylated primer is adjusted so that the biotin binding capacity of the streptavidin coated microtiter wells is not exceeded.

For the minisequencing reactions, two aliquots (or four aliquots for parallel assays) of each biotinylated PCR product are captured in streptavidin-coated microtiter plate wells for 1.5 hours at 37°C. The wells are washed thoroughly and the captured DNA fragment is denatured by treatment with NaOH. The minisequencing primer annealing and extension reactions are carried out simultaneously for 10 min at 50°C in a reaction mixture containing 10 pmol of detection step primer, 0.1 µCi of a [^{3}H]dNTP complementary to the nucleotide to be detected and 0.1 unit of a thermostable DNA polymerase in 50 µl of DNA polymerase buffer. After washing the wells, the primer is released by denaturation with NaOH, and the eluted [^{3}H] is measured in a liquid scintillation counter. The primer sequences and the PCR conditions for the described applications are given in the respective references cited below. A detailed and generally appplicable protocol for the solid-phase minisequencing method is given by Syvänen et al., 1993[13].

RESULTS AND DISCUSSION

Principle of the method

In the solid-phase minisequencing method a DNA fragment spanning the site of the variable nucleotide is first amplified using one biotinylated and one unbiotinylated PCR primer. The

PCR product carrying a biotin residue at the 5' end of its strands is captured on an avidin-coated solid support and denatured. The nucleotides at the variable or mutant site in the immobilized DNA strand are then identified by two separate primer extension reactions, in which a single labelled deoxynucleoside triphosphate (dNTP) is incorporated by a DNA polymerase (Figure 1). In our standard format of the assay [³H]dNTPs serve as labels and streptavidin-coated microtiter plates are used as the solid support[13].

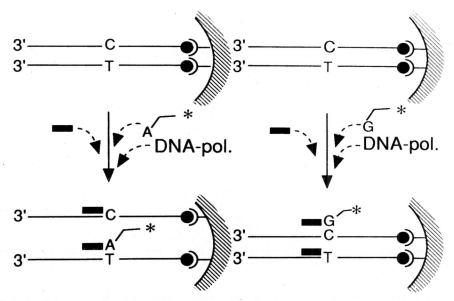

Figure 1. Principle of the solid-phase minisequencing reaction exemplified by the detection of the nucleotides A and G.

The result of the assay are numeric cpm-values expressing the amount of [³H]dNTP incorporated in the minisequencing reactions. The ratio between the cpm-values obtained in the minisequencing assay (the R-value) reflects directly the ratio between the two sequences in the original sample (Figure 2). The method is highly sensitive allowing quantitative determination of one sequence present as a minority of less than 1 % of a sample. Therefore the dynamic range for the quantitative analysis is wide, spanning five orders of magnitude.

Because the two sequences differ from each other by a single nucleotide, they are amplified with equal efficiency during PCR, and the R-value obtained in the minisequencing assay is not affected by the amount of template present in the reaction (Figure 3). Consequently the quantitative analysis can be performed irrespective of whether the PCR process is in its exponential or plateau phase.

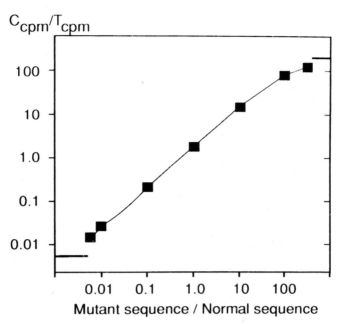

Figure 2. Solid-phase minisequencing standard curve prepared by analysing mixtures of two 63-mer oligonucleotides differing from each other at one nucleotide in the mitochondrial tRNA^{Leu} gene[15]. The mean C_{cpm}/T_{cpm} ratio obtained in three parallel minisequencing reactions is plotted as a function of the original ratio between two oligonucleotides in the mixtures. The horizontal bars indicate the C_{cpm}/T_{cpm} ratio obtained when the two oligonucleotides were analysed separately.

The original ratio between the two sequences in the amplified sample is calculated from the R-value by taking into account the specific activities of the [³H]dNTPs used. If a nucleotide at a polymorphic site is followed by one (or more) identical nucleotide, one (or more) additional nucleotide will become incorporated in the minisequencing reaction, which obviously will affect the R-value. Both these factors are known in advance, and can be corrected for. Alternatively, instead of calculating the initial ratio between the two sequences from the specific activities of the [³H]dNTPs used, the ratio between the two sequences can be determined by comparing the obtained R-value with a standard curve prepared with mixtures of known amounts of the corresponding two sequences. The use of a standard curve corrects for the specific activities and number of [³H]dNTPs incorporated, and in addition, also for any possible (small) misincorporation of [³H]dNTP by the DNA polymerase, which may be significant when a sequence present as a small minority of a sample is to be quantified. Depending on the application, we use human genomic DNA of known genotypes[8] or synthetic oligonucleotides14 as standards for quantification of DNA sequences.

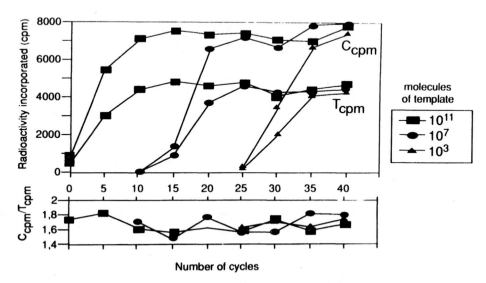

Figure 3. Result of the solid-phase minisequencing assay obtained at different PCR cycles and amounts of template. Mixtures of equal amounts (10^3, 10^7 or 10^{11} molecules) of the same oligonucleotides as in Figure 2 were analyzed. The upper panel shows the cpm-values obtained in the minisequencing assay at different PCR cycles, and the lower panel shows the corresponding C_{cpm}/T_{cpm} ratios.

Determination of allele frequencies by quantitative analysis of pooled DNA samples

We have developed a system for forensic DNA typing, in which a panel of 12 single nucleotide variations giving rise to biallelic sequence polymorphism is analysed by the solid-phase minisequencing method[13]. The statistical interpretation of the result of forensic and paternity testing analyses requires information on the allele frequencies of the analysed markers in each particular population. To rapidly obtain this information in the Finnish population, we utilized the quantitative nature of the solid-phase minisequencing method to determine the allele frequencies of the markers by analysing large pooled DNA samples representing hundreds of individuals. The ratio between the two sequences at each polymorphic locus in the pooled DNA samples is equivalent to the allele frequencies in the population. Table 1 shows as an example the result obtained when the allele frequencies of a polymorphism in the PROS1 gene on chromosome 13 were determined. As with this marker, a good correlation between the allele frequencies determined from the pooled samples and by analysis of about 50 individual samples were observed for each marker[13].

Sample[1]	[³H]dNTP incorporated (cpm)		A_{cpm}/G_{cpm}	Allele frequency[3]	
	A-reaction	G-reaction		A-allele	G-allele
Pool 1390	2,750	1,280	2.14	0.59	0.41
Pool 860	2,240	1,048	2.15	0.59	0.41
Pool 920	2,510	1,190	2.11	0.58	0.42
Control (AA)	2,100	52	40	-	-
Control (GG)	96	1,930	0.050	-	-
Control (AG)	2,480	1,660	1.49	-	-
no DNA	64	39	-	-	-

Table 1. Example of the determination of the allele frequencies of a polymorphism in the PROS1 gene by quantitative analysis of pooled DNA samples. [1] The figure gives the number of individuals in each pool. AA, GG and AG indicate the genotypes of the individual control samples. [2] Mean values of five (pools) or two (individual controls) parallel assays. The specific activities of [³H]dATP and [³H]dGTP were 58 and 32 Ci/mmol, respectively. [3] The allele frequencies determined from 50 individual samples were 0.61 (A) and 0.39 (G).

Analogously, we have determined the frequency of carriers of the recessively inherited disorder aspartylglucosaminuria (AGU) in Finland by determining the frequency of the mutant AGU allele by quantitative analysis of the large pooled DNA samples[8]. The mutant AGU-allele was expected to be present as a small minority (about 1 %) in the DNA pools. Therefore it was essential to use a standard curve constructed with mixtures of known amounts of DNA from a patient homozygous for the AGU-mutation and normal genomic DNA for the interpretation of the result of the quantitative analysis to be accurate.

Detection of heteroplasmic point mutations of the mitochondrial DNA

A typical feature of diseases caused by point mutations in the mitochondrial DNA is, in addition to the fact that they are maternally inherited, that the tissues of the patients often contain both mutant and normal DNA. This phenomenon is called heteroplasmy. A correlation between the degree of heteroplasmy and the severity of the mitochondrial disorders has been suggested. The solid-phase minisequencing is particularly useful for detecting the heteroplasmic mitochondrial mutations because it allows both identification and quantification of the mutation in the same assay. Using the solid-phase minisequencing method we have observed a correlation between the degree of heteroplasmy of a mutation in the mitochondrial tRNA[Lys] gene and the severity of the MERRF-disease (myoclonus epilepsy with ragged-red-fiber) in a large Belgian family[14], and between the amount of mutation in the mitochondrial tRNALeu gene and the severity of the MELAS-disease (mitochondrial encephalomyopathy, lactic acidosis and stroke-like episodes) in a Finnish family[15].

Determination of gene copy numbers

Based on the solid-phase minisequencing method, we developed a less technically demanding alternative than the widely used fluorescence in situ hybridization techniques for the determination of the copy number of human genes. The copy number of a marker gene, aspartylglucosaminidase (AGA), located at chromosome 4qter was determined by this method in three patients with either deletions or duplications involving the distal region of chromosome 4q16. An equal amount of DNA from a patient homozygous for a mutation in the AGA gene was added to the DNA samples to be analyzed. The relative amount of normal sequence in the combined samples was determined by the solid-phase minisequencing method. The expected proportion of normal DNA in samples with two copies of the normal AGA gene is 0.5. In cases where one copy of the AGA gene has been deleted, the ratio should be 0.33 and in case of a duplication the ratio should be 0.6. The expected result was obtained in the experiment (Table 2) showing that the solid-phase minisequencing method is feasible for the determination of monosomies, trisomies and loss of heterozygosity, provided that a DNA standard containing a suitable polymorphism is available.

Caryotype of sample	$C_{cpm}/[C_{cpm}+G_{cpm}]$	Deduced AGA gene copy number
46, XY, -4, +der (4) t(4;12) (q31.3;pl2.2)mat	0.28-0.33[1]	1
46,XX,del (4) (q33)	0.26-0.32[1]	1
46,XX,-21,+der(21) t(4;21) (q28;p13)mat	0.63-0.70[1]	3
Controls	0.50-0.54[2]	2

Table 2. Determination of the copy number of the aspartylglucosaminidase gene (the data are from Laan et al. 16) [1] Range of variation of five parallel assays. [2] Normal range for heterozygote AGU carriers [8].

Identification of mixed samples

The R-values obtained when individual genomic DNA samples are analyzed for a variable nucleotide by the solid-phase minisequencing method fall into three distinct categories that unequivocally define the genotype of the sample. R-values falling outside these three categories that normally differ from each other by a factor of ten, are an indication of the presence of contaminating DNA in the sample. The ability of the method to identify a mixed sample is an advantage in forensic analyses, where stain samples may contain DNA from several individuals, as well as in prenatal diagnosis, where placental biopsy samples may contain contaminating maternal DNA. Obviously, also PCR contaminations will be identified by aberrant R-values. In a recent study that we undertook to develop a method for preimplantation diagnosis by combining whole genome amplification (PEP, primer extension preamplification[17]) of single blastomere cells with PCR and solid-phase minisequencing, the quantitative nature of the solid-phase minisequencing method turned out to be highly informative. It allowed us to notice that preferential amplification of one allele occurs at heterozygous loci during PEP, which is a potential problem in preimplantation diagnosis of dominantly inherited disorders and of recessive disorders caused by compound heterozygous mutations[18].

PITFALLS AND TROUBLESHOOTING

Because [³H]dNTPs that have low specific activities are used as detectable groups in the solid-phase minisequencing method, it is important that the PCR amplification has been efficient. About one tenth of the PCR product should be clearly visible on an agarose gel stained with ethidium bromide.

An obvious prerequisite for complete capture of the biotinylated PCR product in the streptavidin-coated microtiter plate wells is that the primer has been efficiently biotinylated (80-90 %) during its synthesis. Normally the biotinylated primer can be used without further purification, but if necessary, biotinylated oligonucleotides can be purified from unbiotinylated oligonucleotides by HPLC, PAGE or with the aid of disposable ion exchange chromatography columns manufactured for this purpose.

The biotin binding capacity of the microtiter plate wells sets an upper limit for the amount of biotinylated PCR product (and excess of biotinylated primer) that can be present during the capturing reaction. The amount of biotinylated primer during PCR and the amount of PCR product analyzed must be adjusted accordingly.

It is important for the specificity of the minisequencing reaction that all dNTPs from the PCR are completely removed by the washing steps following the capturing reaction. The presence of other dNTPs than the intended [³H]dNTP during the minisequencing reaction will cause unspecific extension of the detection step primer.

The conditions for the primer annealing reaction are non-stringent. Therefore the same reaction temperature (50°C) can be used for practically all 20-mer detection step primers. If the primer is extremely A-T rich, this can be compensated for by increasing the length of the primer.

REFERENCES

1. Syvänen A.-C., M. Bengtström, J. Tenhunen and H. Söderlund. Quantification of polymerase chain reaction products by affinity-based hybrid collection. *Nucleic Acids Res*, 16: 11327 (1988).
2. Murphy L.D., C.E. Herzog, J.B. Rudlick, A.T. Fojo and S.E. Bates. Use of polymerase chain reaction in the quantitation of mdr-1 gene expression. *Biochemistry*, 29: 10351(1990).
3. Noonan K.E., C. Beck, T.A. Holzmayer, J.E. Chin, J.S. Wunder, I.L. Andrulis, A.F. Gazdar, C.L. Willman, B. Griffith, D.D. von Hoff and I.B. Robinson. Quantitative analysis of MDR1 (multidrug resistance) gene expression in human tumors by polymerase chain reaction. *Proc Natl Acad Sci USA*, 87: 7160 (1990).
4. Chelly J., J.-C. Kaplan, P. Maire, S. Gautron and A. Kahn. Transcription of the dystrophin gene in human muscle and non-muscle tissues. *Nature*, 333: 858 (1988).
5. Wang A.M., M.V. Doyle and D.F. Mark. Quantitation of mRNA by the polymerase chain reaction. *Proc Natl Acad Sci USA*, 86: 9717 (1989).

6. Gilliland G., S. Perrin, K. Blanchard and H.F. Bunn. Analysis of cytokine mRNA and DNA: detection and quantitation by competitive polymerase chain reaction. *Proc Natl Acad Sci USA*, 87: 2725 (1990).

7. Syvänen A.-C., K. Aalto-Setälä, L. Harju, K. Kontula and H. Söderlund. A primer-guided nucleotide incorporation assay in the genotyping of apolipoprotein E. *Genomics*, 8: 684 (1990).

8. Syvänen A.-C., E. Ikonen, T. Manninen, M. Bengtström, H. Söderlund, P. Aula and L. Peltonen. Convenient and quantitative detection of the frequency of a mutant allele using solid-phase minisequencing: Application to aspartylglucosaminuria in Finland. *Genomics*, 12: 590 (1992).

9. Ikonen E., T. Manninen, L. Peltonen and A.-C. Syvänen. Quantitative determination of rare mRNA species by PCR and solid-phase minisequencing. *PCR Methods Appl*, 1: 234 (1992).

10. Dieffenbach C.W., T.M.J. Lowe and G.S. Dveksler. General concepts for PCR primer design. *PCR Methods Appl*, 3: S30 (1992).

11. Higuchi R. Simple and rapid preparation of samples for PCR. *In:* PCR technology. Principles and applications. Ed. H.A. Ehrlich. Stockton, New York, p31 (1989).

12. Innis M.A. and D.H. Gelfand. Optimization of PCRs. *In:* PCR protocols: A guide to methods and applications. Eds. M.A. Innis, D.H. Gelfand, J.J. Sninsky and T.J. White. Academic Press, INC. Harcourt Brace Jovanovich, Publishers, San Diego, p3 (1990).

13. Syvänen A.-C., A. Sajantila and M. Lukka. Identification of individuals by analysis of biallelic DNA markers using PCR and solid-phase minisequencing. *Am J Hum Genet*, 52: 46 (1993).

14. Suomalainen A., P. Kollmann, J.-N. Octave, H. Söderlund and A.-C. Syvänen. Quantification of mitochondrial DNA carrying the tRNA$_{8344}^{Lys}$ point mutation in myoclonus epilepsy and ragged-fiber disease. *Eur J Hum Genet*, 1: 88 (1993).

15. Suomalainen A., A. Majander, H. Pihko, L. Peltonen and A.-C. Syvänen. Quantification of tRNA$_{3243}^{Leu}$ point mutation of mitochondrial DNA in MELAS patients and its effect on mitochondrial transcription. *Hum Mol Genet*, 2: 525 (1993).

16. Laan M., K. Grön-Virta, A. Salo, P. Aula, L. Peltonen, A. Palotie and A.-C. Syvänen. Solid-phase minisequencing confirmed by FISH analysis in determination of gene copy number. *Hum Genet*, 96: 275 (1995).

17. Zhang L., K. Cui, K. Achmitt, R. Hubert, W. Navidi and N. Arnheim. Whole genome amplification from a single cell: Implications for genetic analysis. *Proc Natl Acad Sci USA*, 89: 5847 (1992).

18. Paunio T., I. Reima and A.-C. Syvänen. Preimplantation diagnosis by whole-genome amplification, PCR amplification, and solid-phase minisequencing of blastomere DNA. *Clin Chem*, 42: 1382 (1996).

BIOIMAGE ANALYSERS: APPLICATION FOR RIBOZYME KINETICS

Claus Stefan Vörtler and Klara Birikh

Arbeitsgruppe Eckstein, Max Planck Institut für Experimentelle Medizin, Herrmann-Rein-Str. 3, 37075 Göttingen, Germany

INTRODUCTION

BioImage Analyser, PhosphorImager, MolecularImager and others are trade names for a series of instruments, which all make use of a technology developed in the 1980s for the detection and quantification of radioactive molecules: the storage phosphor technology. Its biggest advantages are (i) the detection of radioactive material e.g. polyacrylamide gels with a >10x higher sensitivity and a bigger dynamic region than conventional X-ray film and (ii) the instant quantification of the radioactive molecules separated in the gel with a precision comparable to liquid scintillation counting (LSC)[1]. These features make it very well suited for the kinetic characterisation of ribozymes (catalytically active RNA).

From the variety of ribozymes isolated since RNA catalysis was first described by Altman[2] and Cech[3] one of the smallest is the hammerhead ribozyme. It is composed of only approximately 30 nucleotides and is able to induce the site-specific cleavage of a phospho-diester bond. Not surprisingly, this small ribozyme has attracted considerable interest. It been extensively studied in order to better understand the structure-function relationship of the nucleotide sequence, but also it offers a potential method for gene therapy through the inhibition of gene expression.

MATERIALS AND METHODS

Oligonucleotides were prepared by automated chemical synthesis using phosphoramidites from PerSeptive Biosystems. Deprotection and purification of oligo-ribonucleotides was as described previously[4]. The randomised dN_{10} was synthesised using an equimolar mixture of all four phosphoramidites and purified by denaturing PAGE as described for the oligoribonucleotides. Plasmid pFEL2B, containing the human AChE cDNA 1.6-kb fragment 550-2149, inserted into pGEM7ZF(+)[5], was used for *in vitro* transcription with T7 RNA polymerase. T7 RNA polymerase was purified from the overproducer *Escherichia coli* BL21/

Modern Applications of DNA Amplification Techniques
Edited by Lassner *et al.*, Plenum Press, New York, 1997

pAR1219 kindly supplied by F. W. Studier (Brookhaven). [γ-^{32}P]GTP and [α-^{32}P]UTP were from Amersham. Ribonucleoside triphosphates, restriction endonuclease *Dde*I and its 10x buffer, *E. coli* RNase H, and bulk tRNA from brewer's yeast were from Boehringer Mannheim, restriction endonuclease *Bst*NI and its 10x buffer were from New England Biolabs. Stop mix contained 7M urea, 50mM EDTA, 0.05% Bromophenol Blue, and 0.05% Xylene Cyanol. Computer analysis of the RNA secondary structure was performed using the MFold program (Genetics Computer Group).

IN VITRO TRANSCRIPTION

Plasmid pFEL2B was linearized by the *Hind*III digestion, extracted with phenol-chloroform and precipitated with ethanol. The pellet was dissolved in water and stored at -20°C. For mapping reactions, in vitro transcription was carried out in the presence of [γ-^{32}P]GTP to obtain 5'-labelled transcript. The reaction mixture containing 6μg of linear plasmid DNA, 40mM Tris-HCl (pH8.0), 12mM MgCl$_2$, 10mM DTT, 0.5mM each ribonucleoside triphosphates, 250μCi [γ-^{32}P]GTP and 1000 units of T7 RNA polymerase in a total volume of 100μl, was incubated at 37°C for 1 h. Non-incorporated triphosphates were removed by centrifugation with Microconcentrators-50 (Amicon). After three rounds of concentration from 200 to approximately 10μl the material retained on the membrane was collected and its volume adjusted to 40μl with water. The concentration was measured by UV absorption and the plasmid DNA concentration subtracted to obtain the transcript concentration. Internally labeled in vitro transcript was used for the ribozyme kinetics with reaction conditions as described above, except that the reaction was usually scaled down by half and 50μCi [α-^{32}P] GTP were added.

MAPPING OF ACCESSIBLE SITES

Each mapping reaction contained 2 pmol of 5'-labeled mRNA transcript, 20 mM Tris-HCl (pH7.5), 20mM KCl, 10mM MgCl$_2$, 0.1mM EDTA, 0.1mM DTT, 5% glycerol, 1nmol of dN10, and 0.1 units of RNase H in a total volume of 15μl. All the components of the reaction were combined except for dN10, glycerol, and RNase H and the volume was adjusted to 10μl. The mixture was incubated for 15 min at 37°C. In a separate tube, 1 nmol of dN10 in aqueous glycerol (15%, 5μl) was denatured for 1 min at 75°C and immediately transferred into the reaction mixture. RNase H was added and the complete reaction mixture incubated at 37°C for 15 min. A control sample was prepared which did not contain dN10. The digested RNA was precipitated with ethanol and taken up in 10μl of water and 10μl of Stop mix. Aliquots of 5μl were separated by electrophoresis in a 5% denaturing PAGE (40 cm long) overnight at 15 W. Dried gels were analysed using the BAS2000 imaging system (Fuji) to determine the positions of RNase H cleaving the mRNA.

RIBOZYME KINETICS

Kinetic constants k$_{react}$ and K$_m$ for cleavage of the *in vitro* transcripts were determined under single turnover conditions from the Eadie-Hofstee plot according to the following equation:

$$-\ln(\text{FracS})/t = k_{obs} = k_{cat} - K_m \, (k_{obs}/\text{RE})$$

where RE is ribozyme concentration and FracS is the fraction of the initial substrate not cleaved at a given time[6]. For cleavage kinetics stock solutions of 1mM ribozyme in 50mM Tris-HCl, pH7.5 were denatured at 75°C for 1 min, followed by 10 min at 37°C, MgCl$_2$ addition to a final concentration of 10mM and further incubation for 5 min at 37°C. Parallel to this substrate RNA was preincubated for 15 min at 37°C in 50mM Tris-HCl, pH7.5, 10mM MgCl$_2$. Reactions were started by the addition of ribozyme (with increasing final concentration of 0.02 to 1.5µM) to the substrate (at 10nM). After 40 to 80 min incubation at 37°C, the reactions were stopped by addition of 10µl stop mix and analysed by 6% denaturing PAGE, 4 h at 50 W, using 40 cm gels. The remaining fractions of the substrate (FracS) over time were determined using a BAS2000 BioImager system (Fuji) with the TINA software package (Raytest) by determining the amount of radioactivity corresponding to the product as well as to the uncleaved substrate. Observed rate constants were calculated from this data using the Kaleidagraph software (Synergy Software).

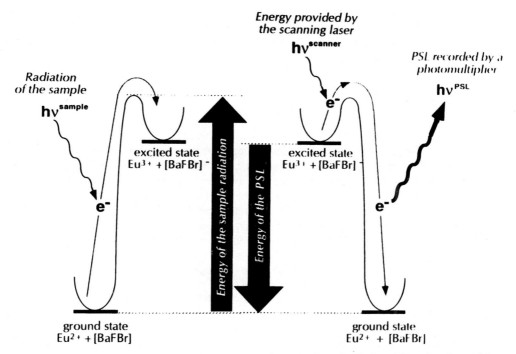

Figure 1. Energy diagram of a storage phosphor screen during recording of radiation (left) and read-out of the stored information (right).

A BRIEF DESCRIPTION OF THE BIOIMAGER TECHNOLOGY

Devices based on storage phosphor technology make use of complex metal-ion-crystals in which electrons from the central metal ion are shuttled between an excited and the ground state[1,7]. During an autoradiographic experiment electrons get excited by energy resulting from radiation hitting a screen called imaging plate or IP, which replaces con-ventional X-ray film and contains a thin layer of such crystals. Once excited, the electrons can just return if additional energy is supplied. More and more electrons will thus accumulate in the excited state during the experiment, being a measure of the amount of radioactive material (of usually constant radiation energy output) and the exposure time. Supplying the needed amount of energy during a subsequent scanning process releases the trapped electrons to the ground state and frees the energy of the excited state as luminescence. This photostimulated luminescence, for short PSL, is measured and its intensity distribution over the scanned IP displayed (Figure 1).

Since a long time interval passes between excitation and return to ground state, the process is defined as phosphoresence. Hence the crystal, the screen as well as the technology carries „storage phosphor" in their names. Conventional Eu[BaFBr] based IP's can be used to record different isotopes, although ^{32}P and ^{33}P together with ^{125}I are most favourable due to their higher radiation energy. Still in our experience the BAS-III screens designed for ^{32}P-detection can be used to detect ^{35}S in an DNA sequencing protocol (exposure time of 6-12h) and even ^{14}C for aminoacylation kinetics (exposure time of >12h). For a reduction of exposure time and background (see troubleshooting, below) in case of the weaker emitting isotopes, special adapted IP's can be obtained from manufacturers. Even ^{3}H, the weakest emitting isotope used in the biochemical research, is detectable with such adapted screens.

RESULTS AND DISCUSSION

A comprehensive characterisation of a ribozyme used for the inhibition of gene expression requires the following steps: cleavage of a short non-structured substrate and cleavage of a long target RNA fragment *in vitro*, proof of ribozyme function in cultured cells, in an animal model and, finally, in patients. *In vitro* experiments are an important initial part of ribozyme research, allowing the screening of a large number of ribozymes to select the most efficient for further *in vivo* investigation. Several reports underline the correlation of *in vitro* and *in vivo* activity of ribozymes[8,9,10], antisense RNAs[11] and antisense oligodeoxynucleotides[12]. In studies where such a correlation has not been observed, the *in vitro* activities were usually obtained under non-physiological conditions[13,14]. Depending on the assay condition chosen, *in vitro* results have predictive power for the performance *in vivo*.

The methodology for ribozyme kinetics using short substrates is very well developed. However, long RNA molecules are usually cleaved orders of magnitude slower and it poses experimental difficulties to observe multiple turnover cleavage. A BioImage analyser is a very valuable tool in this context, since standard kinetic assays consist out of a dPAGE step to separate cleavage-product from uncleaved substrate, followed by autoradiographic detection and quantification of both. An IP with its higher sensitivity (in the order of 10-100 times)[1,7] compared to conventional X-ray film is a big advantage, enabling product detection even in the case of bad cleaving ribozymes. Furthermore, quantification is faster compared to the alternative procedure relying on localisation, excision and LSC of the gel pieces. With an IP the data used for image generation can be directly applied for quantification in a precision comparable to LSC. Since image and quantitation data are digital they can directly loaded into software used for graphic processing and kinetic calculations, respectively. Here we use a BioImage Analyser to localise accessible sites in a long mRNA substrate and to characterise the ribozymes directed against these sites.

Single turnover kinetics (when each ribozyme molecule participates in not more than one round of cleavage) are used in some laboratories for quantitation of long substrate cleavage. For these kinetics substantial excess of the ribozyme is used during the reaction. Cleavage proceeds as a first order reaction of ribozyme-substrate-complex decay, therefore cleavage rate only depends on the ribozyme concentration as it affects the fraction of bound substrate. A set of observed rate constants obtained at various ribozyme concentrations allows the determination of k_{cat} and K_m using the Michaelis-Menten equation or the linear representations of Michaelis-Menten graphs, termed Eadie-Hofstee plots. From these *in vitro* studies it is important to select ribozymes with low K_m values for the later application, as ribozymes have to function *in vivo* efficiently at comparatively low concentration.

An additional important point is the association of the ribozyme with the target site, which is usually the rate limiting factor for long substrate cleavage (as a result of the complex secondary and tertiary interactions within the folded substrate RNA). It is a important pre-requisite for the inhibition of gene expression by ribozymes to identify such RNA sites supporting optimal ribozyme annealing. Our group was interested in developing ribozymes to interfere with expression of the human AChE gene. Such interference may be a start to develop therapies of neurodegenerative diseases resulting from AChE overexpression. In many studies computer folding programs are used to find single-stranded regions which are considered as potential cleavage sites for ribozymes[15]. We used such a program, MFold, to target three ribozymes, RZ1259, RZ1305 and RZ1386, against the human AChE transcript. Each of these ribozymes cleaved its respective short substrate with reasonable efficiency (data not shown). However, only RZ 1259 cleaved the 1.6 kb AChE transcript, the remaining two were inactive (Table 1). The possible reason lies in the folding programs, such as MFold, which only identify a few out of the whole set of possible alternative RNA structures with similar free energies[16]. Besides, in a long RNA molecule locally stable structures may exist which kinetically block the RNA folding into the global free energy minimum structure[15]. Thus, an alternative experimental-based approach would be desirable to identify ribozyme accessible regions in a long RNA molecule.

A completely randomised oligodeoxynucleotide (dN_{10}) in conjunction with RNase H allowed us to find sites accessible for oligonucleotide binding to the human AChE transcript (RNase H cleaves just where RNA:DNA hybrids are formed - no hybridisation, no cleavage). Cleavage products were directly detected after dPAGE of reaction mixtures containing dN_{10} and transcript in a molar ratio of 500:1. As the total number of individual decamers in dN_{10} is 4^{10} (about 10^6) the sequence-defined dN_{10} to transcript molar ratio was actually much lower, approximately 1:2000, resulting in a limited cleavage-ladder. Six accessible sites were identified according to the gel-mobility of the produced RNA fragments. Three of them had five potential cleavage codons, either CUC or UUC, in their vicinity. Ribozymes direct against these codons followed the assumption that the sites accessible for oligodeoxynucleotides should also be accessible for ribozymes. All were found to be much more active in cleaving the transcript than to those selected on the basis of the secondary structure prediction (Table 1). In fact, the best of the five ribozymes, RZ1888, proved to be ~150 times more active than the best of the MFold-directed ones. It is noteworthy that RZ1888 is even a better catalyst (with a k_{react}/K_m value of $6.8 \times 10^5 \, M^{-1}min^{-1}$ in the case of long substrate) than those found previously for other long substrate targets[6,17,18] (values between 0.2-$0.8 \times 10^5 \, M^{-1}min^{-1}$). Cleavage rates k_{react} for the long substrate transcript were generally one to two orders of magnitude lower than those for the short substrate (data not shown). This is in agreement with earlier findings[6,17,18] and indicates a different rate-limiting step for the two substrates. The three ribozymes (RZ1888, RZ1194, RZ1173) with the highest cleavage efficiencies k_{react}/K_m have considerably lower K_m values than RZ1870 and RZ1212, which also cleave less efficiently (Table 1). Given that the cleavage rate k_{react} for both are very similar, is the K_m value indicative of a good catalyst .

As already mentioned, these kind of kinetic analyses can also be done using X-ray films in conjunction with LSC. Laser Densitometery using directly the X-ray films as another alternative suffers from several draw-backs: a lower sensitivity, a non-linear response curve to radiation and variability depending on the developing conditions. The latter two points are particularly problematic for quantification purposes. Furthermore: the dynamic region, the region in which a low intensity signal is still detectable and a high intensity signal causes no over exposition, is five orders of magnitude in the case of an IP[7] - one to two orders more then for X-ray film. This bigger region pays off for the characterisation of poorly cleaving ribozymes, since always the ratio between cleaved (weak signal) and uncleaved substrate (strong signal) is needed.

The storage phosphor technology has already served several purposes since its invention. The primary aim was to digitalize medical diagnostic radiography. IPs are now used throughout the biochemical research fields and are even found in area-detectors of crystallography. As outlined here,   their quantification features are also of considerable value for ribozyme research.

Ribozyme	K_{react} $(10^{-3}min^{-1})$	K_m (nm)	K_{react}/K_m $(10^4 M^{-1}min^{-1})$	Annealing arms to the core sequence -CUGAAGAGGCCGAAAGGCCGAA-
MFOLD:				
RZ1259	1.9+/-0.5	420+/-10	0.45	CAGAAAA-**core**-ACGAGCC
RZ1305	n.d.	n.d.	n.d.	CUGAUGA-**core**-ACUCGUU
RZ1386	n.d.	n.d.	n.d.	AAUGCAG-**core**-ACCACAG
dN10/RNase				
RZ1888	15+/-1	22+/-6	68.2	CCUCGUC-**core**-AGCGUGU
RZ1194	4.4+/-0.4	40+/-10	11.0	CGUUCAU-**core**-AGGGCCU
RZ1173	3.8+/-0.1	54+/-7	7.0	UGUCACU-**core**-GGAAGU
RZ1870	6.1+/-0.8	160+/-10	3.7	UGGCGCU-**core**-AGCAAUU
RZ1212	2.2+/-0.7	250+/-30	0.9	GCCGUGC-**core**-AGUCUCC

Table 1. Catalytic constants of ribozymes directed against selected AChE mRNA target sites using the M-fold program as well as the dN_{10} / RNase H assay.

TROUBLESHOOTING

Using a BioImager in the daily lab routine

Imaging Plates are designed to have a life time of at least 1000 exposition/scanning events. This is just the lower limit: one of our IPs has been in use since 1989 and after probably ten times as many expositions still working accurate enough to do kinetics with it. To reach such a long lasting performance it is important to avoid damage to the sensitive IP surface. The biggest danger is moisture, since the protective sheet covering the IP-surface provides just protection against dirt or radioactive contamination. It helps at best against an accidental spill of water on the surface (if immediately removed). Any prolonged contact to moisture, like direct contact to wet polyacrylamide gels, bears the danger of water diffusing through this layer and damaging the crystal lattice underneath causing yellowish spots on the originally white surface. Partial or complete loss of sensitivity towards radiation is the result and loss of image information from parts of a scanned IP indicate this problem. There is no remedy against it, the IP is destroyed. Even if no visible damage appears on the plate, the sensitivity can be affected - which leads to wrong quantification results. Drying polyacrylamide gels prior to IP exposition is the easiest way to avoid problems. As alternative, we also expose wet gels after covering them carefully with one or two layers of conventional saran wrap and one layer of a special hostaplan-film (obtained from Raytest). The latter provides an impermeable barrier against moisture and IP exposition of the wet gel for up to 10-12h seems safe .

Since the image is recorded by trapping electrons in an excited state, there is a steady fading of the image observable if the IP is not directly scanned after exposition. Up to 20% loss in the first three hours have been reported[7], which can be avoided by rapid scanning at the end of exposure. Another problem is a uniform greyish background caused by cosmic and natural radiation hitting the IP. It is visible after long-time exposures of days rather than hours or scanning at highest sensitivity levels. It can be a problem if a very weak signal with the intensity-level of this background needs to be detected. This background can be reduced using a lead container during IP exposure. To erase the background accumulated during storage of the IP, a short 1 min erasure prior to the start of exposition is advisable.

REFERENCES

1. Amemiya Y. and J. Miyahara. Imaging plate illuminates many fields. *Nature,* 336:89 (1988).
2. Guerrier-Takada C., K. Gardiner, T. Marsh, N. Pace, S. Altman. The RNA moiety of Ribonuclease P is the catalytic subunit of the enzyme. *Cell,* 35:849-857 (1983).
3. Kruger K., P. Grabowski, A. Zaug, J. Sands, D. Gottschling, T. Cech. Self-splicing RNA: Autoexcision and autocyclization of the ribosomal RNA intervening sequence of Tetrahymena. *Cell,* 31: 147(1982).
4. Tuschl T., M.M.P. Ng, W. Pieken, F. Benseler, F. Eckstein. Importance of exocyclic base functional groups of central core guanosine for hammerhead ribozyme activity. *Biochemistry,* 32: 11658-11668 (1993).
5. Soreq H., R. Ben-Aziz, C. Prody, S. Seidman, A. Gnatt, L. Neville, J. Lieman-Hurwitz, E. Lev-Lehman, D. Ginzberg, Y. Lapidot-Lifson, H. Zakut. Molecular cloning and construction of the

coding region for human acetylcholinesterase reveals a G+C-rich attenuating structure. *Proc Natl Acad Sci USA,* 87: 9688-9692 (1990).

6. Heidenreich O., F. Benseler, A. Fahrenholz, F. Eckstein. High activity and stability of hammerhead ribozymes containing 2´-modified pyrimidine nucleosides and phosphorothioates. *J Biol Chem,* 269: 2131-2138 (1994).

7. Johnston R., S. Pickett, D.Barker. Autoradiography using storage phosphor technology. *Electrophoresis,* 11: 360 (1990).

8. Lieber A. and M. Strauss. Selection of efficient cleavage sites in target RNAs by using a ribozyme expression library. *Mol Cell Biol,* 15: 540-551(1995).

9. Lieber A. and M. Kay. Adenovirus-mediated expression of ribozymes in mice. *J Virology,* 70: 3153-3158 (1996).

10. Sun L.Q., D. Warrilow, L. Wang, C. Witherington, J. Macpherson, G. Symonds. Ribozyme-mediated suppression of Moloney murine leukemia virus and human-immunodeficiency-virus type I replication in permissive cell lines. *PNAS,* 91: 9715-9719 (1994).

11. Rittner K., C. Burmester, G. Sczakiel. *In vitro* selection of fast-hybridising and effective antisense RNAs directed against the human immunodeficiency virus type 1. *Nucl Acids Res,* 21: 1381-1387 (1993)

12. Ho S.P., D.H.O. Britton, B.A. Stone, D.L. Behrens, L.M. Leffet, F.W. Hobbs, J.A. Miller, G.L. Trainor. Potent antisense oligonucleotides to the human multidrug resistance-1 mRNA are rationally selected by mapping RNA-accessible sites with oligonucleotide libraries. *Nucl Acids Res,* 24: 1901-1907 (1996)

13. Crisell P., S. Thompson, W. James. Inhibition of HIV-1 replication by ribozymes that show poor activity *in vitro. Nucl Acids Res,* 21: 5251-5255 (1993).

14. Dropulic B. and K. Jeang. Intracellular susceptibility to ribozymes in a tethered substrate-ribozyme provirus model is not predicted by secondary structures of human immunodeficiency virus type 1 RNAs *in vitro. Antisense Res Develop,* 4: 217-221 (1994).

15. Christoffersen R.E., J. McSwiggen, D. Konings. Application of computational technologies to ribozyme biotechnology products. *J Molec Structure (Theochem),* 311: 273-284 (1994).

16. Zuker M. and P. Stiegler. Optimal computer folding of large RNA sequences using thermodynamics and auxiliary information. *Nucl Acids Res,* 9: 133-148 (1981).

17. Hendrix C., J. Anne, B. Joris, A. Van Aerschot, P. Herdewijn. Selection of hammerhead ribozyme for optimum cleavage of interleukin 6 mRNA. *Biochem J,* 314: 655-661 (1996).

18. Marschall P. Inhibition der Geneexpression mit Ribozymen. Dissertation, Universität Göttingen, Göttingen (1996).

FIRST APPROACHES TO QUANTITATE MDR1-MESSENGER RNA BY *IN CELL* PCR

Dirk Lassner[1], Joerg Milde[1], Matthias Ladusch[2], Karl Droessler[2] and Harald Remke[1]

[1]Institute of Clinical Chemistry and Pathobiochemistry, University Leipzig, Liebigstraße 16, 04103 Leipzig. [2]Institute of Zoology, University Leipzig, Talstraße 33, 04103, Germany

INTRODUCTION

The MDR1 gene encodes the glycoprotein P-170 which is mostly the molecular base of multidrug resistance (MDR) providing a channel for an active efflux of cytostatics out of the cells. The MDR1 phenotype is often induced by applied chemotherapeutics and the detection of MDR1-mRNA within leukocytes of leukemic patients represents a good marker for this resistance[1]. Polymerase chain reaction (PCR) following reverse transcription (RT) of MDR1-mRNA provides an extremely sensitive method to detect low levels of transcripts or transcripts from a small number of cells.

In general the detection of MDR1-mRNA is performed by PCR related methods. The level of gene expression measured quantitatively by different methods are related to a definite amount of isolated RNA[2-9], but there is no correlation of copy numbers to the percentage of cells expressing MDR1-mRNA. The total RNA from a population containing a small number of mRNA high-expressing cells contains the same copy number as a population with 100% low-expressing cells. For treatment of leukemic patients by chemotherapy it will be necessary to estimate the number of cells expressing high or low levels of MDR1-mRNA.

In cell PCR was used for detection of MDR1-mRNA on cellular level in leukocytes. This method combines the intracellular amplification of mRNA by RT-PCR in fixed cells followed by flow cytometric analysis [10,11]. The cells of different cell lines (CCRF CEM, CCRF ADR 10000, HL60)[12] were fixed with paraformaldehyde. The *in cell* PCR was performed with a definite number of cells using FITC-labeled primers. The fluorescence of cells were measured by flow cytometry after various numbers of PCR cycles (20, 25, 30, 35, 40). The cell fluorescence after amplification corresponds to the amout of MDR1-mRNA transcripts. The expression of MDR1 gene in the resistant cell lines (CCRF CEM, CCRF ADR10.000) were

Modern Applications of DNA Amplification Techniques
Edited by Lassner *et al.*, Plenum Press, New York, 1997

compared to MDR1-mRNA of the non-expressing, promyeloic cell line HL-60. By introduction of difference fluorescence the gene expression of leukemic patients can be evaluated relatively. By *in-cell* PCR it will be firstly possible to analyze nucleic acids on single-cell level in a great population of different cells.

MATERIALS AND METHODS

Cell Culture

The T-lymphoblastoid cell line CCRF CEM and its adriamycin resistant mutant CCRF ADR 5000 was originally selected by Gekeler et al.[12]. The adriamycin high-resistant cell line CCRF ADR 10.000 was selected accordingly from the cell line CCRF ADR 5000.

All cells were grown as stationary suspension cultures in RPMI 1640 supplemented with 10% heat-inactivated FCS, 1% tylosine tartrate, and 0.16mg/ml gentamycin in a humidified atmosphere of 95% air and 5% CO_2 at 37°C. They were subcultured every 3 days. The resistant subline CCRF-ADR 10000 is adapted to the continous presence of 10μg/ml adriamycin (Lilly, Italy).

Cell Harvest and Fixation

The fixation was performed according to Haase et al.[10]. Cells were harvested by centrifugation (600g x 5 min), washed twice in PBS-CMF and fixed for 30 min at ambient temperature in freshly prepared 4 % (w/v) paraformaldehyde in PBS-CMF. After fixation the cells were dehydrated by addition of 2 volumes of 100 % (v/v) ethanol.

RT-PCR

Prior RT reaction fixed cells were pelleted, resuspended in PBS-CMF and allowed to rehydrate for at least 20 min. The cells were counted and aliquots of 10^5 cells were transferred to MicroAmp tubes (Perkin Elmer, USA). After centrifugation (1,000 g x 5 min) the supernatant was discarded and cells were resuspended in 16μl reverse transcriptase mixture containing 2μl 5x AMV reverse transcriptase buffer (250mM Tris-HCl, pH 8.3; 250 mM KCl, 50mM MgCl$_2$, 50mM DTT, 2.5mM spermidine), 1μl 10 mM dNTPs and 200ng of downstream primer NP2[3]. The suspended cells were heated to 65°C for 10 min and then cooled on ice. After addition of 4μl enzyme mix (10 units AMV reverse transcriptase (Promega, USA), 20 units RNase Inhibitor (Pharmacia, Sweden), 2μl 5 x AMV reverse transcriptase buffer) the reaction was incubated for at least 1 hour at 42°C.

After RT reaction the tubes were cooled on ice, 30μl of a PCR mix containing 3μl 10 x PCR buffer (500 mM KCl, 100 mM, pH 8.3, 15 mM MgCl$_2$, 0.01 % Gelatine), 200 μM dNTPs, 200 ng upstream primer NP1[3] and 2 units Taq DNA polymerase (Perkin Elmer, USA) were added and the amplification was performed by "hot start" technique (60°C)[13] over 40 cycles. A fluorescein-labeled primer NP1 was used for following analysis of cells by flow cytometry.

Analysis of amplified DNA by agarose gel electrophoresis

10 μl of the PCR sample were run on 2 % (w/v) Qualex Gold (AGS Heidelberg, Germany) agarose gel in 1 x TAE buffer supplemented with ethidium bromide (4 μl/100 ml buffer).

Flow Cytometry

After *in vitro* amplification cells were pelleted, resuspended and washed twice in PBS/2 % BSA. The cytofluorographic analysis of cells after *in cell* PCR was performed on a FACScan flow cytometer (Becton Dickinson, USA). The data from 10^4 cells per sample were collected and stored using LYSIS II software. Cell debris and doublets were gated out and green fluorescence was measured by optical bandpass filter (525 nm) after laser excitation at 488 nm.

RESULTS AND DISCUSSION

Principle of method

In cell PCR combines the amplification of nucleic acids inside cells followed by detection of PCR products by flow cytometry, agarose gel electrophoresis or *in situ* hybridisation [10,11,14,15].

Instead of isolated RNA whole cells were added to a PCR assay. Amplification was performed up to 40 cycles (Figure 1). For following analysis of cells by flow cytometry it is necessary to prevent the destruction of cellular structure by thermal cycling. Representative results were achieved by fixation of the cells with 4 % paraformaldehyde and ethanol[10].

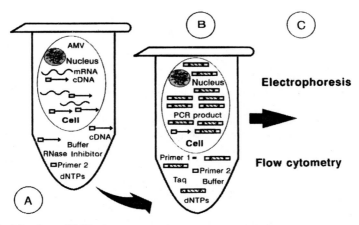

Figure 1. Principle of *in cell* PCR: **A.** Reverse transcription of specific mRNA. **B.** Amplification of cDNA. **C.** Detection of amplified DNA by agarose gel electrophoresis or flow cytometry.

The use of *in cell* PCR was extended by reverse transcription of MDR1-mRNA by specific priming (Figure 1A) and the subsequent amplification of retained MDR1-cDNA by addition of a PCR mix to complete RT assay (Figure 1B).

The amplified DNA is fluorescence labeled within the cells using a fluorescent primer in the amplification step. The analysis of washed cells after *in cell* PCR by flow cytometry (Figure 1C) allows the detection of fluorescent cells and the measurement of fluorescence intensities of individual cells.

Detection of MDR1-mRNA in paraformaldehyde fixed cells

The method of *in cell* PCR was used to amplify MDR1-mRNA in fixed cells of drug resistant cell lines CCRF CEM, CCRF ADR 10.000 and the promyelogeneous cell line HL-60. The optimal cell number was defined by different dilutions of MDR1-mRNA high-expressing cell line ADR 10000 subjected to in-cell PCR. The amplification was performed over 40 cycles and the PCR products were analysed by agarose gel electrophoresis (Figure 2).

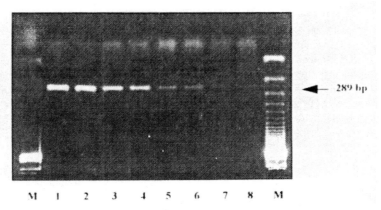

Figure 2. Analysis of *in cell* PCR within different numbers of paraformaldehyde fixed cells (CCRF ADR 10.000) by agarose gel electrophoresis. M: marker (123 bp-DNA ladder, Gibco), lane 1-8: cell number (100.000, 50.000, 10.000, 5.000, 1.000, 500).

Fluorescence detection of intracellular PCR product

The amplified DNA within cells could be visualised by conventional fluorescence microscopy after incorporation of one fluorescent PCR primer. We used a fluorescein-amidite labeled primer which is incorporated by amplification and not during RT reaction.

The following analysis of fluorescence by flow cytometry confirmed that the majority of cells were fluorescent. The whole cells were gated from debris and doublets and the fluorescence intensity was measured logarithmically (Figure 3A). The green fluorescence of the cell lines with different expression of MDR1 gene (CCRF CEM - low expression, CCRF ADR 10.000 - high expression, HL-60 - no expression)[16] was detected after 40 cycles of amplification (Figure 3B).

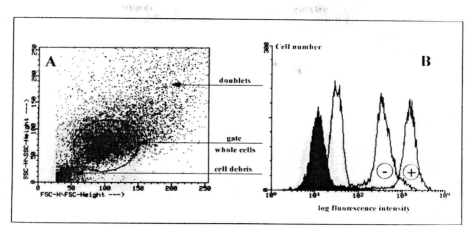

Figure 3. Flow cytometry of cells with amplified MDR1-DNA. **A.** Forward scatter (FSC) /Side scatter (SSC) analysis of cells after *in-cell* PCR and gate of whole cells. **B.** Green fluorescence of different cells after *in cell* amplification with fluorescent primers (dark peak - fixed cells, light peak - fixed cells after *in cell* PCR without fluorescent primers, "-": fixed cells without amplified MDR1-DNA (HL-60); "+": fixed cells with amplified MDR1-DNA (CCRF ADR 10000).

Kinetic analysis of *in cell* PCR by flow cytometry

The number of target molecules of the polymerase chain reaction can be accurately determined by measuring the molar concentration of a product which accumulates in consecutive cycles[17]. The concentration of the product can be calculated from a labeling substance incorporated by amplification[18].

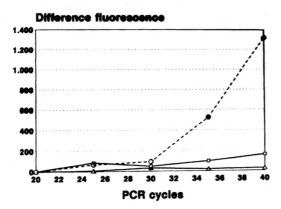

Figure 4. Kinetic analysis of *in cell* PCR by comparison of difference fluorescence. Difference fluorescence is the difference between green fluorescence at indicated PCR cycle and green fluorescence at PCR cycle 20 of corresponding cell line measured by flow cytometry after in-cell PCR (i - CCRF ADR 10000, -CCRF CEM, D- HL-60).

The fluorescence of cells subjected to *in cell* PCR with fluorescent primers results of non-specific background and fluorescence of incorporated primers in generated PCR products which are located within cells.

We amplified the intracellular MDR1-mRNA over 40 cycles for kinetic analysis of *in cell*

PCR by flow cytometry and measured the green fluorescence of cells at different number of cycles (10, 20, 25, 30, 35, 40). An exponential increase of fluorescence intensity started at PCR cycle 20 (Figure 4). Before this point the fluorescence results predominantly from non-specific incorporation of fluorescent primers.

Cell line/Patient	Parameter	Fluorescence at cycle 35	Ratio of fluorescences CCRF ADR 10.000 to
CCRF ADR 10000	Cell fluorescence	957	1.00
high expression	Difference Fluorescence	524	1.00
CCRF CEM	Cell fluorescence	516	1.85
low expression	Difference fluorescence	99	5.82
HL 60	Cell fluorescence	361	2.65
no expression	Difference fluorescence	61	8.59
positive patient	Cell fluorescence	693	1.38
high expression	Difference Fluorescence	333	1.57
negative patient	Cell fluorescence	643	1.48
no expression	Difference fluorescence	56	9.29

Table 1. Estimation of expression of MDR1-mRNA by *in cell* PCR. The proportion of fluorescences of cell line CCRF ADR 10.000 to fluorescences of other samples was used for validation of MDR1-mRNA expression.

For determination of specific fluorescence the fluorescence value at PCR cycle 20 was substracted from the cell fluorescence at indicated PCR cycle. This amount is called difference fluorescence. The comparison of difference fluorescence allowed a quantitative evaluation of specific accumulation of FITC-labeled PCR product within cells by *in cell* PCR (Table 1).

The proportion of difference fluorescence of MDR1-mRNA high-expressing cell line CCRF ADR 10.000 to difference fluorescence of comparable sample was used for quantitative validation of MDR1 gene expression. A value of this proportion about 1-2 is indicating high-expressing cells. Values over 5 to 9 are determined in low or non-expressing cells (Table 1).

Determination of difference fluorescence at different number of consecutive PCR cycles allowed the estimation of amplification efficiency[18] of *in cell* PCR. The amplification efficiency of the high-expressing cell line CCRF ADR 10.000 was 0.24 between cycle 20 to 40. This value means a 300-fold increase of cell fluorescence over 30 cycles. Comparable data were achieved by analysis of *in situ* PCR[19,20].

DISCUSSION

Treatment of leukemic patients with leukocytes expressing MDR1-mRNA with cytostatics could be dangerous for the patient. This treatment could result in a selection of high resistant cell population which will not respond to applied chemotherapy. Early and reliable diagnosis of percentage of blood cells with MDR1 phenotype is essential for successful treatment with modified scheme of cytostatics.

We described detection of intracellular MDR1-mRNA by *in cell* PCR and flow cytometry. First approach for quantitation of MDR1 gene expression in cell lines and patient samples by *in cell* PCR was performed by application of mathematical analysis.

Beside characterisation of drug resistant cells of leukemic patients MDR1-mRNA and the P-glycoprotein reached new importance as second stem cell marker of hematopoesis[21].

In cell PCR for MDR1-mRNA firstly described in this paper will offer a new tool for fast determination of MDR1-mRNA positive cells in the whole blood of leukemic patients or for analysis of transplantation material. Combination of very sensitive RT-PCR and flow cytometry will enable clincans to analyse a lot of samples within a short time period.

TROUBLESHOOTING

The *in cell* PCR is a combination of two important diagnostic methods with different aims (Table 2). The consideration of specific requirements and different sources of errors of both methods is essential for optimal application of *in cell* PCR technology.

Flow cytometry	Polymerase chain reaction
well defined labelling of cells	specific products
good light scatter properties of cells	high amplification efficiencies
high fluorescence intensities	detection of nucleic acids sequences
distinct different fluorescence peaks	estimation of copy numbers
estimation of cell numbers	

Table 2. Comparison of specific demands to flow cytometry and PCR as diagnostic tools

Fixation of cells has influence on accessment of PCR components to target sequences and on stability of cell structure through thermal cycling during PCR. Cross-linking fixatives like formalin or paraformaldehyde hinder successful amplification without considerable pre-treatment[22]. We achieved best performance of *in cell* PCR by fixation of cells with 4 % paraformaldehyde and subsequent dehydration with ethanol. The application of non-crosslinking fixatives (PermeaFix)[15] results in narrow peaks but in limited amplification of MDR1-mRNA.

The number of cells subjected to PCR assay is critical. Overloading of sample (more than 10^5 cells) leads to reduced number of PCR products or failure of amplification.

Direct incorporation of fluorescent primers increases the non-specific cell fluorescence and complicates the distinct differentation of cells within amplified DNA from cells without specific PCR product[11]. This problem makes it difficult to analyse a mixed population of marker-positive and -negative cells. An *in situ* hybridisation with a fluorescent probe after *in cell* PCR without FITC-labeled primer can circumvent this obstacle[10,14,15].

Retention of amplified product in the cells is essential for following analysis of *in cell* PCR by flow cytometry. We amplified a small fragment (289 bp) of MDR1 gene which was membrane-permeant. This results in a loss of cell fluorescence and, more critically, offers the possibility of carry-over contamination during amplification process[26]. MDR1-mRNA negative cells became positive by incorporation of newly generated PCR product and the measured fluorescence was not corresponding to native situation within this cell prior *in cell* PCR. Larger PCR products (650 bp and 1,400 bp) were retained inside cells but were very difficult to amplifiy[16,22].

The single-tube RT-PCR with AMV reverse transcriptase and Taq DNA polymerase is a powerful tool to detect mRNA transcripts *in vitro*[24,25]. *Thermus thermophilus* (rTth) DNA polymerase acts both as reverse transcriptase and DNA polymerase and is optimal for single-tube RT-PCR[14,15,26]. Application of rTth DNA polymerase instead of combined system with AMV reverse transcriptase and Taq DNA polymerase for *in cell* PCR leads to more narrow peaks in flow cytometry but no exponential amplification was measured.

ACKNOWLEDGEMENT

We wish to thank Dr. T. Koehler and Dr. S. Leiblein (Leipzig) for the cell material and patient samples which were used in this study. We are very grateful to Dr. V. Gekeler (Konstanz) and Dr. H. Diddens (Lübeck) for providing us with drug resistant cell lines.

REFERENCES

1. Goldstein L.J., I. Pastan and M.M.Gottesman. Multidrug resistance in human cancer. *Crit Rev Oncol/Hematol*, 12: 243 - 253 (1992)
2. Noonan K.E., C. Beck, T.A. Holzmayer, J.E. Chin, J.S. Wunder, I.L Andrulis, A.F. Gazdar and I.B. Roninson. Quantitative analysis of mdr1 (multidrug resistance) gene expression in human tumors by polymerase chain reaction. *Proc Natl Acad Sci USA*, 87:7160-7164 (1990).
3. Köhler T., D Laßner, A-K. Rost, S. Leiblein and H. Remke. Polymerase chain reaction related approaches to quantitate absolute levels of mRNA coding for the multidrug resistance-associated protein and P-glycoprotein. In "Advances in blood disorder", R. Pieters, G.J.L Kaspers, A.J.P. Veermann (eds.). Harwood Academic Publishers, Bershire (1996).
4. de Kant E., C.F. Rochlitz and R. Herrmann. Gene expression analysis by a competitive and differential PCR with antisense competitors. *Biotechniques*, 17: 934-942 (1995).
5. Bremer S., T Hoof, M Wilke, R. Busche, B. Scholte, J.R. Riordan, G. Mass and B. Tümmler. Quantitative expression pattern of multidrug-resistance P-glycoprotein and differentially spliced cystic fibrosis transmembran-conductance regulator mRNA transcripts in human epithelia. *Eur J Biochem*, 206:137-149 (1992).
6. Murphy L.D., C.E. Herzog, J.B. Rudick, A.T. Fojo and S.E. Bates. Use of polymerase chain reaction in the quantitation of mdr-1 gene expression. *Biochemistry*, 29:10351-10356 (1990).
7. Hoof T., J.R. Riordan and B. Tümmler. Quantitation of mdr1 transcript by PCR a tool for monitoring drug resistance in cancer chemotherapy. *Anal Biochem*, 196: 161-169 (1991).
8. Futscher B.W., L.L. Blake, J.H. Gerlach, T.M. Grogan and W.S. Dalton. Quantitative polymerase chain reaction analysis of mdr1 mRNA in multiple myeloma cell lines and clinical specimens. *Anal Biochem*, 213: 414-421 (1993).
9. Grünebach F., E.U. Griese and K. Schumacher. Competitive nested polymerase chain reaction for

quantification of human mdr1 gene expression. *J Cancer Res Clin Oncol,* 120:539-544 (1994).

10. Haase A.T., E.F. Retzel and K.A. Staskus. Amplification and detection of lentiviral DNA inside cells. *Proc Natl Acad Sci USA,* 87 :4971-4975 (1990).

11. Embleton M.J., G. Gorochov, P.T. Jones and G. Winter. *In-cell* PCR of mRNA: amplifying and linking the rearranged immunglobulin heavy and light chain V-gene within single cells. *Nucl Acids Res,* 1992; 20:3831-3837.

12. Gekeler V., S. Wegner and H. Probst. Mdr1/p-glycoprotein gene segments analyzed from various human leukemic cell lines exhibiting different multidrug resistance profiles. *Biochem Biophys Res Comm,* 169:796-802 (1990).

13. Ehrlich H.A., D. Gelfand and J.J.Sninsky. Recent advances in the polymerase chain reaction, *Science,* 252:1643-1651 (1991).

14. Patterson B.K., M. Till, P. Otto, C. Goolsby, M.R. Furtado, M.J. McBride and S. Wolinsky. Detection of HIV-1 DNA and messenger RNA in individual cells by PCR-driven *in situ* hybridization and flow cytometry. *Science,* 260:976-979 (1993).

15. Patterson B.K., C. Goolsby, V. Hodara, K. Lohmann and Steven Wolinsky. Detection of CD4+ T cells harboring human immunodeficiency virus type 1 DNA by flow cytometry using simultaneous immunotyping and PCR driven *in situ* hybridization: evidence of epitope masking of the CD4 cell surface molecule *in vivo. J Virol,* 69:4316-4322 (1995).

16. Lassner D., unpublished data.

17. Wiesner R.J., J.C. Rüegg and I. Morano. Counting target molecules by exponential polymerase chain reaction: copy number of mitochondrial DNA in rat tissues. *Biochem Biophys Res Commun,* 183:553-559 (1992).

18. Wiesner R.J. Direct quantification of picomolecular concentrations of mrRNAs by mathematical analysis of a reverse transcription/exponential polymerase chain reaction assay. *Nucl Acids Res,* 20:5863-5864 (1992).

19. Nuovo G.J. PCR *in-situ* hybridization: protocols and applications. Raven Press, New York (1992).

20. Embretson J., M. Zupancic, J. Beneke, M. Till, S. Wolinsky, J.L. Ribas, A. Burke and A.T. Haase. Analysis of human immunodeficiency virus-infected tissues by amplification and *in situ* hybridization reveals latent and permissive infections at single cell resolution. *Proc Natl Acad Sci USA,* 90:357-361 (1993).

21. Chaudary P.M. and I. Robinson. Expression and activity of P-glycoprotein, a multidrug efflux pump, in human hematopoietic stem cells. *Cell,* 66:85-94 (1991).

22. Nuovo G.J., F. Gallery, R. Hom, P. MacConnell and W. Bloch. Importance of different varia- bles for enhancing *in situ* detection of PCR-amplified DNA. *PCR Meth Appl,* 2:305-312 (1993).

23. Longo M.C., M.S. Berninger and J.L. Hartely. Use of uracil DNA glycosylase to control carry-over contamination in PCRs. *Gene,* 93:125-128 (1990).

24. Aatsinki J.T., J.T. Lakkakorpi, E.M. Pietilä and H.J. Rajaniemi. A coupled one-step reverse transcription PCR procedure of generation of full-length open reading frames. *BioTechniques,* 16: 282-288 (1994).

25. Pfeffer U., E. Fecarotta and G. Vidali. Efficient one-tube RT-PCR amplification of rare transcripts using short sequence-specific reverse transcription primers. *BioTechniques,* 18:204- 206 (1995).

26. Myers T.W. and D.H. Gelfand. Reverse transcription and DNA amplification by a *Thermus thermophilus* DNA polymerase. *Biochemistry,* 30: 7661- 7666 (1991).

APPLICATION OF *IN SITU-PCR* FOR THE DETECTION OF INTRACELLULAR mRNAs

Volker Uhlmann, Eilhard Mix and Arndt Rolfs

Department of Neurology, University of Rostock, Gehlsheimerstr. 20, 18147 Rostock, Germany

INTRODUCTION

In the last decade, polymerase chain reaction (PCR) has probably had a greater impact on molecular biology than any other technique. This highly sensitive technique is capable of amplifying a single copy of a gene to a level detectable by gel electrophoresis and Southern blotting. As conventional PCR generally requires cell or tissue destruction in order to isolate the nucleic acid template, it is not possible to visualize or localize the amplicon within a single cell. *In situ* hybridization (ISH) permits localization of nucleic acid sequences with high specificity at the level of individual cells. Whilst preserving the cell morphology, its application is limited by the low detection sensitivity of approximately 20 copies per cell at its best.

The lack of sensitivity of standard ISH has led several research groups to attempt to combine the methodologies of solution phase PCR and ISH. The new but challenging *in situ*-PCR methodology consists of a „marriage" of both techniques combining the advantages of ISH, i.e. high specificity and localization at the level of an individual cell, with the advantage of PCR, i.e. high sensitivity.

One of the earliest publications describing *in situ*-PCR was by Haase et al. [1]. Lentiviral DNA was amplified in infected cells *in situ*, and the amplicon was subsequently detected by ISH. In the meantime several variations of the situ PCR methodology have been independently developed for different applications.

In our studies we have demonstrated the expression of the specific human dopamine D2 receptor mRNA in peripheral blood lymphocytes using a slide-based nonisotopic *in situ* reverse transcription (RT) assay. There are five subtypes of dopamine receptors (D1- D5) expressed in human brain, predominantly in the nucleus accumbens and caudate nucleus. Since the 1990s there is evidence to assume that most of the dopamine receptors are also expressed in peripheral blood lymphocytes (PBL) as demonstrated by RT and subsequent PCR [2]. The function and regulation of the transcripts in PBL is still unclear. The development of severe central nervous system (CNS) diseases like schizophrenia and Parkinson's disease are closely related to the D2 receptor. Additionally there may exist a relationship between the

Modern Applications of DNA Amplification Techniques
Edited by Lassner *et al.*, Plenum Press, New York, 1997

level of dopamine D2 receptor transcripts and inflammatory diseases of the CNS.

MATERIAL AND METHODS

Although there is a rising number of publications in the field of *in situ*-PCR, it must be emphasized that there exists no single protocol to be recommended exclusively so far: all have merits and pitfalls and should be optimised by the researchers for their own. It is necessary to analyze the *in situ*-PCR protocols as a series of different stages which are shown in the following flow chart.

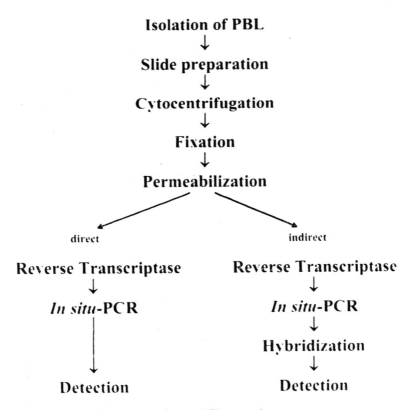

In Situ-RT-PCR

Isolation of PBL
↓
Slide preparation
↓
Cytocentrifugation
↓
Fixation
↓
Permeabilization

direct   indirect

Reverse Transcriptase   **Reverse Transcriptase**
↓   ↓
In situ-PCR   *In situ*-PCR
                                              ↓
                                        Hybridization
↓   ↓
Detection   **Detection**

Figure 1. Flow chart of the different stages of *in situ*-PCR protocols

Isolation of PBL

PBL were isolated from freshly drawn EDTA-blood by Histopaque density gradient separation. The cells were washed twice in standard culture medium (RPMI) adjusted to 2.5 x 10^5 cells/ml of RPMI containing 10% fetal calf serum (FCS) and cultured for up to 24h in humidified atmosphere with 6% CO_2. Alternatively one can isolate PBL by lysing the erythrocytes with an appropriate buffer.

Slide Preparation

To perform slide-based *in situ*-PCR, the cells have to be immobilized on glass slides coated with binding reagents such as APES (Aminopropyl-Triethoxy-Silane), Denhardt's solution, poly-L-lysine or Elmer's glue. The binding depends on electrostatic or hydrophobic interactions between chemical groups of the agent and cell surface constituents [3].

For our preparation procedure silane-coated slides from Perkin Elmer are used, which have been designed to fit to other parts of the equipment to create a non-evaporating reaction chamber.

Charged, as well as hydrophobic groups on the oligonucleotides and the Taq polymerase might react with reagents like APES. Therefore, it is advisable to coat the slides with a solution of bovine serum albumin (BSA) prior to the PCR reaction [4]. For increasing the amplification efficacy it is in some cases helpful to add BSA also to the PCR reaction mixture (final concentration up to 0.1%).

Cytocentrifugation

The conditions for achieving good cell preparations on precoated slides are straightforward: 200µl of the cell suspension (approx. $5x10^4$ cells in 10% FCS medium) were cytocentrifuged at 400 x g for 5 min. 150µl of the supernatant were carefully discarded. With a second centrifugation step (1,000 x g) for 1 min, the cells were rapidly dried.

Fixation

Prior to *in situ* amplification the cytoskeleton of the cell must be fixed. This step creates a micro-environment within the cell which facilitates entry of all PCR components and minimizes leakage of amplified product. The „combination" of fixation and subsequent permeabilization causes the major practical problems of this technique. In general terms, longer fixation times seem to entail a longer digestion step. At this point it is important to stress that the different specimens (cryostat tissue sections, cell smears, fixed cell suspensions and paraffin embedded tissue) and of course different *in situ* applications such as *in situ*-PCR and *in situ* RT-PCR necessitate different fixation procedures.

We tested different types of fixatives. At first the cells were fixed in 10% neutral buffered formaldehyde (NBF, 10% formalin in PBS, pH 7.2) for 20 hours. This resulted in the destruction of cell surface epitopes preventing two colour staining experiments. We found the use of acetone (for 15 min at ambient temperature) as a non cross-linking fixative preferable. In recent studies we tested Permeafix™ (Ortho Diagnostics; 60 min at ambient temperature), which is a good compromise between the crosslinking and precipitating agents. Moreover it is also a permeabilizing reagent [5].

Crosslinking	Non-crosslinking
10% Neutral buffered formaldehyde (NBF) over night	Acetone for 15 min
4% Neutral buffered paraformaldeyde (NBF) over night	Permeafix™ for 1 hour
Glutaraldehyde	Ethanol
Bouin's solution (formalin with picric acid)	Methanol/acetic acic (3:1)
Zenker's solution (formalin with heavy metals)	Streck's fixative
	Heat fixation

Table 1. Different fixatives used for *in situ*-PCR

Aldehyde fixatives such as formaldehyde and paraformaldehyde can react with the histones of the nucleic acid template resulting in crosslinks within the cell. The majority of these crosslinks are protein-protein interactions with some crosslinking of DNA to proteins but no crosslinking of DNA to DNA [3]. No crosslinking has been observed with RNA, which is important for all assays including an RT-step. However, even after several washes with phosphate buffered saline (PBS, pH 7.2) aldehyde groups are still present and can continue forming crosslinks. In addition, crosslinked histone proteins prevent the access of Taq polymerase to its nucleic acid target. Interestingly, BSA is not crosslinked to DNA by formalin such as other proteins [6]. This is worth emphasizing, since BSA may play an important role in increasing the PCR amplification rate and in blocking the inhibitory influence of APES. In particular, tissue samples need to be fixed with formaldehyde fixatives to prevent the loss of cell morphology. This crosslinking net also helps to prevent outward diffusion of the amplicon.

The use of highly crosslinking fixatives such as glutaraldehyde or mercuric chloride modified formalin should be avoided, as they render the tissue virtually impermeable to the PCR components even when extremely harsh pretreatments have been used [7]. Picric acid based fixatives also result in tisssue specimens which are difficult to amplify.

Since our primary interest focuses on the mRNA detection in single suspended cells, precipitating fixatives such as acetone, ethanol or mixture of ethanol and acetic acids are useable. The advantage of these fixatives is that there is no crosslinking interaction which results in a more efficient PCR amplification. It has been reported that non-crosslinking fixatives allow penetration with no or little pretreatment. However, in tissues - in contrast to our acetone fixed cells - there seems to be only a slight retention of the final product in the cells and after finishing the thermal cycling process the morphology looks in most cases rather poor [7,8]. The problem of diffusion can be reduced by incorporating biotinylated nucleotides to generate bulky and less diffusable amplicons [9]. Other possible improvements have been made by Chiu et al. [10] and Patterson et al. [11] who have successfully used multiple primer pairs in combination with non-crosslinking fixatives. Recently, the usage of Permeafix™ (Ortho Diagnostics) was reported [5]. This product is a mixture of a fixative that preserves antigenicity and of a detergent-based permeabilizing agent that allows entry of PCR reagents.

Finally, when choosing a suitable fixative for *in situ* amplification short fixation times and protease digestion times and the conflicting advantages of the different fixatives have to be considered.

Permeabilization

Cell permeabilization is carried out in order to allow penetration of the PCR components into the cytoplasm or into the nucleus. Permeabilization of specimens has been achieved employing proteases at different concentrations [1,9,12,13,14,15]. Commonly used proteases are proteinase K (10- 500µg/ml), trypsin (2mg/ml), pepsin (2mg/ml), pronase (2mg/ml) and trypsinogen (2mg/ml).

In our first experiments 300µg/ml proteinase K (in 0.1 M Tris, pH 7.2) were used for 15 min on 10% NBF fixed lymphocytes. Since mRNA is located in the cytoplasm and does not react with formaldehyde, even gentler digestion conditions are practicable. In order to maintain antigenicity of the cell surface of PBL permeabilization and fixation can be carried out in acetone for 15 min and with Permafix™ for 1 hour at ambient temperature.

Excessive protease digestion leads to „DNA-repair- mechanisms" or „endogenous priming"[16,17]. Interestingly, hydrochloric acid treatment increases the *in situ*-PCR amplification [18,19]. It appears that hydrochloric acid may partially solubilize the crosslinked histones enabling an easier access of the PCR components to the target DNA. A combined protease and detergent treatment (e.g. NP40, Triton X100, Tween 20 etc.) often leads to better results because they partially permeabilize in addition to the proteases. However, even no usage of any digestion pretreatment in *in situ*-PCR has been reported [20].

Depending on the enzyme for immunohistochemical detection - alkaline phosphatase or horseradish peroxidase - it is recommended to block endogenous enzyme prior to detection. A solution of 0.1% sodium azide and 0.3% hydrogen peroxide for 10 min is sufficient to quench endogenous peroxidase in PBL [21].

In all cases the optimum concentration of protease for particular tissues or cells has to be determined empirically by titration with the aim to preserve a good morphology on one hand and to avoid the diffusion of the final PCR product out of the cells on the other.

Reverse Transcription

In the actual literature plenty of different protocols for *in situ* RT-PCR have been published [11,20,22,23,24,25,26]. For initiating the reverse transcriptase step, antisense oligonucleotide primers anneal complementary to their specific site of the mRNA and the reverse transcriptase enzyme such as that derived from the *Mouse Moloney Leukemia Virus* (MMLV) or *Avian Myelobastosis Virus* (AMV) polymerizes the cDNA that is in the following step the target for the Taq polymerase catalyzed PCR amplification process.

A typical 10µl reaction mixture for the RT-reaction consists of: 2µl 5x reaction buffer (50mM Tris, pH 8.3; 75mM KCl), 5mM MgCl$_2$, 8µl with 3mM DTT, 0.5mM dNTPs, 0.3mM downstream primer, 40 U RNase inhibitor (RNAsin), 200U MMLV (Superscript). After an incubation for 5min at 55°C and 1 hour at 37°C in a moist microtiter plate a fixation step with 100% ethanol follows at room temperature for 2 min in a coplin jar.

As a control for preventing erroneous coamplification of genomic DNA, in some protocols prior to reverse transcription a DNase digestion step is introduced [8,27]. Using special designed primers for exclusive cDNA amplification („junction primers") which overspan the RNA splice junction sites, there seems to be no need to perform such a DNase step. Moreover, incomplete DNase digestion almost leads to undesired „DNA repair mechanism" artefacts which are the major problem in the direct labeling *in situ*-PCR assay (see below).

In most RT-PCR assays two different enzymes (reverse transcriptase and Taq polymerase) and therefore also two different reaction buffers were used. Patterson and coworkers [11] have introduced the usage of the recombinant rTth polymerase which combines the enzymatic activity of a DNA polymerase and a reverse transcriptase and is also active at 72°C. This design simplifies the assay since the RT and PCR reactions can be done without changing the buffer or without adding a chelating buffer [3]. The typical reaction components for such a 50μl assay includes: 10μl 5 x reaction buffer (50mM bicine; 125mM potassium acetate; 40% glycerol, pH 8.2), 4- 6mM manganese acetate, 400μm dNTPs, 1μM upstream and downstream primer, 7.5U rTth polymerase (Perkin Elmer, USA). The cycling conditions for the RT-PCR are 30 min at 60°C followed by a two step PCR (primer annealing and elongation at the same temperature).

During the complete RT-step you have to be aware of RNases. Therefore we recommend to use exclusively solutions and tools which have been treated before use with diethylpyrocarbonate (DEPC) in a final concentration of 0.1%.

In situ-PCR

A special variant of *in situ*-PCR is the direct labelling of the resulting PCR product. This variant omits the hybridization step by incorporating modified labeled oligonucleotides like biotin-dUTP or digoxigenin-dUTP. This procedure has the advantage of being less time-consuming because of the failing of the hybridization step, but is has the disadvantage of more pronounced artefacts since labeled oligonucleotides are incorporated into partially fragmented DNA undergoing "repair"by the Taq DNA polymerase and into non-specific PCR products primed by „endogenous" DNA and cDNA fragments. [28]

Using the direct labeling of the PCR products our 50μl *in situ*-PCR cocktail consists of: 10mM Tris [pH 8.0], 50mM KCl, 5.5mM MgCl$_2$ (take care of the high magnesium concentration), 0.4μM dNTP, 1.0 μM of each primer, 5U Taq polymerase (pay attention to the high concentration of the enzyme), 0.02μM digoxigenin-11-dUTP (Dig-dUTP: dTTP = 1:20). A two step PCR with 25- 30 cycles is applied.

The *in situ*-PCR appears applicable to any gene as long as the oligonucleotide primers have been proven prior to be effective and specific in a standard PCR assay. Owing to mispriming phenomena we preferably used the „hot start" technique which reduces mispriming and primer oligomerization. This technique describes the adding of Taq polymerase to a preheated master mix and the preheated slide (70°C). Nuovo and coworkers [16] have suggested the use of single stranded binding protein (SSB) derived from *E. coli*, which is involved in DNA replication, prevents mispriming and pimer oligomerization and allows a „cold start" technique. The usage of anti-Taq polymerase antibodies seems to have the same effect.

To increase amplification effiency, several groups made use of PCR mixtures with higher concentrations of primers, magnesium and DNA polymerase than employed for standard solution phase PCR [14,16]. The need of relatively high concentrations of these reagents likely reflects sequestration on the slide. This seems to be demonstrable in the presence of BSA, since there is often a clear signal enhancement, although lower Taq polymerase concentrations are used [16].

One of the most critical parameters of PCR is the attainment of correct temperatures for each PCR step. "Thermal lag", i.e. differences in temperatures between the block surface, the glass slide and the PCR mix at each temperature step of the reaction cycle, occurs[21]. Recently equipments have become available which offer in-built slide temperature calibration curves that correct these „thermal lag" phemomena.

25 to 30 cycles have been found to be adequate in most cases of *in situ* amplification [9,12,14,29]. The annealing step is the rate-determining one in PCR. In our experiences a single annealing/ extension step at 55°C for 2 minutes is sufficient [8]. It has been reported that higher numbers of cycles significantly increase the amount of diffused PCR product. Of course this effect depends directly on the size of the final product [9]. The incorporation of biotinylated nucleotides into the amplicon renders the product more bulky and therefore reduces the chance of diffusion of the final product.

It is clear that *in situ* amplification is not as efficient as solution phase PCR. The extent of slide- based amplification over 30 cycles is about 200- 300 fold [8,13,24] assuming an amplification rate of only 1.1-1.2 fold per cycle. The extent of amplification from *in situ*-RT-PCR is difficult to ascertain, because it requires the knowledge of the amount of the labile mRNA within the cell [3].

Primer selection has evolved around two basic stratagies: single primer pairs [9,12,14,15,29] or multiple primer pairs with or without complementary tail [1,10,13,14,15,23,24]. The multiple primer pair approach generates longer and and/or overlapping PCR product with less potential to diffuse from its place of origin. It is reasonable to assume that most of the amplified products may have diffused out of the cell following physical rules. Therefore it is advisable to wash and fix (100% ethanol) the slides immediately after the reaction has finished. We recommend to follow this protocol step even when the bulky hapten molecule digoxigenin was incorporated in the amplicon. Nevertheless, it is an essential control to take an aliquot of the supernatant in order to run an agarose gel. This may be evidence for a positive *in situ*-PCR reaction.

Detection of amplicons

Direct *in situ*-PCR generates amplicons within the cytoplasm that have incorporated digoxigenin labeled nucleotides. For indirect *in situ*-PCR digoxigenin-labeled probes have to be added in a hybridization step. In both cases digoxigenin molecules can be visualized using a conventional two- or three-step enzymatic detection system. Before antibody incubation the slides have to be washed in 2 x SSC at 45°C to remove excess of digoxigenin-11-dUTP. A further preincubation with TBT-buffer is necessary. Three incubation steps follow: (1) monoclonal mouse anti-digoxigenin (Boehringer Mannheim, Germany) (diluted 1:50 in TBT); (2) biotinylated rabbit anti- mouse F(ab')2 (diluted 1:50 in TBT); (3) streptavidin/peroxidase (Dako, Hamburg, Germany) (diluted 1:50 in TBT with 5% blocking reagent, e.g. dried skim milk). Finally aminoethyl-carbazol (AEC)-chromogen is added which yieldes a red signal [18]. Between each step the slides were washed with TBS-buffer. The three step detection system is slightly more sensitive than the one step technique by a higher amplification level of the signal through a second antibody step. All slides were counterstained with haemalaun for approximately 10- 15 min and were mounted in glycerol jelly.

It has been demonstrated [30,31] that the detection limit of labeled RNA or DNA can be increased by using the tyramide signal amplification system (TSA) This is based on the deposition of biotinylated tyramide by the activity of peroxidase. This system is a new method for signal amplification, which boosts the the signals up to a 1.000-fold. The reaction is rapid (about 1-10 min) and results in the covalent deposition of a biotin label. Remarkably, the added biotin tyramide is found to be deposited extremely close to or right at the enzyme and the secondary biotinylated antibody [32].

Controls

The importance of appropiate and adequate controls for *in situ*-PCR can not be overemphasised. These are the controls to be required for direct *in situ*-PCR:

Control	Purpose
Reference control gene e.g. pyruvate deydrogenase (PDH)	- Control of specificity and sensitivity of the used method
DNase digestion	- Abolishment of genomic signals (negative control) - for RT: reduction of false positives
RNase digestion	- Reduction of false positives - for RT: abolishment of RNA signal (negative control)
Omission of primers	- Detection of artefacts related to DNA repair mechanisms
Omission of RT step (only relevant for RT assay!)	- Detection of mispriming
Omission of labeled nucleotides/ labeled primers	- Control of the detection system
Omission of Taq polymerase	- Detection of primer dimerisation/oligomerisation
Mixtures of known positive and negative cells; identification of different cell types by immuno-histochemistry	- for RT: Control of specificity/sensitivity of the method

Table 2. List of different controls to be required for direct *in situ*-PCR

TROUBLESHOOTING

The following table summarizes possible explanations for pitfalls and problems with *in situ*-PCR and suggestive strategies for their avoidance:

Problem	Explanation	What to do
Loss of cells	No or unsufficient coating of the slides	Coat the slides with APES
Evaporation of PCR mixture	No tight seal	Choose semi-automated closed chamber system (Perkin Elmer)
Weak or no signals (false negatives)	- Reagent loss due to non-specific adsorption to the slide	Coat the slides with BSA or increase the amount of manganese/magnesium, primers and Taq
	- Ineffective amplification	- BSA in PCR mixture - Increase numbers of cycles
Signal outside the cells (high background)	Leakage of product/diffusion of product	- Incorporate biotin substituted dNTP's - Increases size of product or use multiple primer pairs or concatamer primer - Fix after PCR with 100% ethanol or paraformaldehyde
Destruction of cell architecture	Harsh digestion conditions or short fixation procedure	Minimize digestion step or prolong fixation procedure
False positive results	Non specific signals „diffusion artefacts"	- „Hot start technique" - Generate more complex PCR products - Reduce PCR cycles - Optimize fixation/permeabilization step
False positive results within the nuclei	Endogenous priming/DNA repair mechanism (a special problem of direct *in situ* PCR!)	- Dideoxy blocking - Super denaturation - DNase digestion or UV cross-linking of the DNA - Use of labeled primers - Use of Stoffel fragment

Table 3. Potential pitfalls and problems with *in situ*-PCR

CONCLUSION

The *in situ*-PCR technology is most useful *in situ*ations where *in situ*-hybridization fails due to low copy numbers. Despite of a few technical problems, *in situ*-(RT) PCR is a powerful technique and provides a rapid method for the localization of cellular mRNA and quantitating positive cells. *In situ*-PCR has a great impact in research and will have it in the diagnostic field in the future. Therefore, it will be necessary to simplify the technique in order to make it available also for routine laboratories.

Our direct *in situ*-PCR protocol was improved by three single main procedures:
(1) Non-crosslinking fixatives like acetone or Permeafix (Ortho Diagnostics) led to better results regarding morphology and antigenicity than the combination of NBF fixation and subsequent proteinase K digestion.
(2) The usage of only one enzyme (rTth polymerase) and only one buffer for the reverse transciption and the following PCR is a rapid and easy alternative in comparison to the two step reaction using two different enzymes, like Taq polymerase and MMLV.
(3) The tyramide signal amplification system has improved the signal intensity of the three step detection method using peroxidase and thereby simplified the detection of positive cells.

For the less time and material consuming direct *in situ*-PCR assay special care has to be taken to avoid the problems of non-specific DNA synthesis.

ACKNOWLEDGMENT

The authors are indebted to J. J. O´Leary/Oxford for outstanding help and a numerous of critical statements and without whom this project would not have been possible.

REFERENCES

1. Haase, A. T., E. F. Retzel and K.A. Staskus. Amplification and detection of lentiviral DNA inside cells, *Proc Natl Acad Sci USA*, 87: 4971- 75 (1990).
2. Takahashi, N., Y. Nagai, S. Ueno, Y. Saeki and T. Yanagihara. Human peripheral blood lymphocytes express D5 dopamine receptor gene and transcribe the two pseudogenes, *FEBS Lett*, 314: 23- 25 (1992).
3. Teo, I. A. and S. Shaunak. Polymerase chain reaction *in situ*: an appraisal of an emerging technique, *Histochem J*, 27: 647- 59 (1995).
4. Yap, E.P. and J. O' D. McGee. Slide PCR: DNA amplification from cell samples on microscopic glass slides, *Nucleic Acids Res*, 19: 4294 (1991).
5. Patterson, B. K., D. Jiyamapa, E. Mayrand, B. Hoff, R. Abramson and P.M. Garcia. Detection of HIV-1 DNA in cells and tissue by fluorescent *in situ* 5'- nuclease assay (FISNA), *Nucleic Acids Res*, 24: 3656- 58 (1996).
6. Solomon, M. J. and A. Varshavsky. Formaldehyde-mediated DNA-protein crosslinking: a probe for *in vivo* chromatin structure, *Proc Natl Acad Sci USA*, 82: 6470- 74 (1985)

7. Lewis, F. A. PCR Manual Perkin Elmer, (1996).

8. Nuovo, G. J. PCR in-situ hybridization: protocols and applications, New York, Raven Press (1992).

9. Komminoth, P., A. A. Long, R. Ray and H. J. Wolf. *In situ* polymerase chain reaction detection of viral DNA, single copy genes and gene rearrangements in cell suspensions and cytospins, *Diagn Mol Pathol*, 1: 85- 87 (1992).

10. Chiu, K. P., S. H. Cohen, D. W. Morris and G.W.Jordan. Intracellular amplification of proviral DNA in tissue sections using the polymerase chain reaction. *J Histochem Cytochem*, 40: 333- 41 (1992).

11. Patterson, B. K., M. Till, P. Otto, C. Goolsby, M. R. Furtado, L. J. McBride and S. M. Wolinsky. Detection of HIV-1 DNA and messenger RNA in individual cells by PCR-driven *in situ* hybridization and flow cytometry. *Science*, 360, 976- 79 (1993).

12. Bagasra, O., S. P. Hauptmann, H.W. Lischner, M. Sachs and R. J. Pommerantz. Detection of human immunodeficiency virus type 1 provirus in mononuclear cells by *in situ* polymerase chain reaction, *New Engl J Med*, 326:1385-91 (1992).

13. Embretson, J., M. Zupanic, T. Beneke, M. Till, S. Wolinsky, J. L. Ribas, A. Burke and A. T. Haase. Analysis of human immunodeficiency virus-infected tissues by amplification and *in situ* hybridization reveals latent and permissive infections at single- cell resolution. *Proc Natl Acad Sci USA*, 90: 357- 61 (1993).

14. Long, A. A., P. Komminoth and H. F. Wolfe. Comparison of indirect and direct *in situ* polymerase chain reaction in cell preparations and tissue sections - Detection of viral DNA, gene rearrangements and chromosomal translocations. *Histochemistry*, 99: 151-62 (1993).

15. Nuovo, G. J., P. MacConnell, A. Forde and P. Delvenne. Detection of human papillomavirus DNA in formalin-fixed tissues by *in situ* hybridization after amplification by polymerase chain reaction. *Am J Pathol*, 139: 847- 54 (1991).

16. Nuovo, G. J., F. Gallery, P. MacConnell and W. Bloch. Importance of different variables for enhancing *in situ* detection of PCR-amplified DNA. *PCR Meth Appl*, 2: 305- 12 (1993).

17. Ansari, B., P. J. Cates , D. Greenstein and P.A. Hall. In-situ end-labeling detects DNA strand breaks in apoptosis and other physiological and pathological states. *J Pathol*, 170: 1- 8 (1993).

18. Bosshoff, C., T. F. Schulz, M. M. Kennedy, A.K. Graham, C. Fisher, A. Thomas, J. O'D. McGee, R. A. Weiss and J. J. O'Leary. Kaposi's sarcoma-associated herpesvirus infects endothelial and spindle cells. *Nature Med*, 1: 1274- 78 (1995).

19. Moran, R., Z. Darzynkiewiecz, L. Staino-Coico and M.R. Melamed, Detection of 5'-bromodeoxyuridine (Br-dURD) incorporation by monoclonal antibodies: role of denaturation step. *Cytochemistry*, 33: 821- 27 (1985)

20. Patel, V. G., A. Shum-Siu, B.W. Heniford, T. J. Wieman and F. J. Hendler, Detection of epidermal growth factor receptor RNA in tissue sections from biopsy specimens using *in situ* polymerase chain reaction. *Am J Pathol*, 144: 7- 14 (1994).

21. O'Leary, J. J., J. Smith, R. J. Landers, A.K. Graham and J. O'D. McGee. *In situ*-PCR/ PCR *in situ* hybridization; practical manual (monograph), Perkin Elmer Cetus/ABI, *European Molecular Biology Workshop Series*, Sept. 1985.

22. Chen, R.H. and S. V. Fuggle. *In situ* cDNA polymerase chain reaction a novel technique for detecting mRNA expression. *Am J Pathol*, 143: 1527- 34 (1993).

23. Embleton, M., G. Gorochov, P. Jones and G. Winter. *In-cell* PCR from mRNA, amplifying and linking and rearranged immunoglobulin heavy and light chain V- genes within single cells. *Nucleic Acids Res*, 20, 3831-37 (1992).

24. Staskus, K., L. Couch, P. Bitterman, E. Retzel, M. Zupancic, J. List and A.T. Haase. *In situ* amplification of visna virus DNA in tissue section reveals a reservoir of latently infected cells.

Microb Pathog, 11: 67- 76 (1991).

25. Kelleher, M. B., D. Galutira, T. D. Duggan and G. J. Nuovo. Progressive multifocal leukoencephalopathy in a patient with Alzheimer's disease. *Diagn Mol Pathol*, 3: 105- 13 (1994).
26. Nuovo, G. J., K. Lidonnici, P. MacConnell and B. Lane. Intracellular localization of polymerase chain reaction (PCR)-amplified hepatitis C cDNA. *Am J Surg Pathol*, 139: 1239- 44 (1993).
27. Staecker, H., M. Cammer, R. Rubinstein and T. van de Water. A procedure for RT-PCR amplification of mRNAs in histological specimens. *BioTechniques*, 16: 76- 80 (1994).
28. Komminoth, P. and A. A. Long. *In situ* polymerase chain reaction, methodology, applications and pathways (Review), PCR Applications Manual (Boehringer Mannheim) (1995).
29. Spann, W., K. Pachmann, H. Zabnienska, A. Pielmeier and B. Emmerich. *In situ* amplification of single copy gene segments in individual cells by polymerase chain reaction. *Infections*, 19: 242-44 (1991).
30. Bobrow, M. N., T. D. Harris, K. J. Shaugnessy and G.J. Litt. Catalyzed reporter deposition, a novel method of signal amplification - application in immunoassays. *J Immunol Meth*, 125: 279-85 (1989).
31. Bobrow, M. N., K.J. Shaugnessy and G.J. Litt. Catalyzed reporter deposition, a novel method of signal amplification (II) - application to membrane immunoassays. *J Immunol Meth*, 137: 103- 12 (1991).
32. TSA™- Indirect amplification for chromogenic and fluorescent immunohistochemistry. Laboratory manual of the manufacturer (Du Pont Nen) (1996).

PSORALEN BIOTIN: A NOVEL REAGENT FOR NON-ENZYMATIC AND SPECIFIC LABELING OF NUCLEIC ACID PROBES AND OLIGONUCLEOTIDES

Robert L. Burghoff[1], Jens Beator[2] and Michael A. Harvey[1]

[1] Schleicher & Schuell Inc., PO Box 2012, Keene, NH 03431, USA.
[2] Schleicher & Schuell GmbH, PO Box 4, 37582 Dassel, Germany

INTRODUCTION

Psoralens are planar tricyclic compounds which bind to both single and double-stranded nucleic acids by hydrophobic interactions with pyrimidine bases. Upon irradiation with long wavelength UV-light between 320 - 400 nm, psoralens form covalent cycloaddition products, preferentially with the bases thymidine and uracil and to a lesser extent also with cytosine[1,2].

We have developed a derivative of hydroxymethyl trioxsalen containing a 15 atom spacer terminating with a biotin molecule. The novel reagent, psoralen biotin, offers a new alternative method for labelling of nucleic acids which is very efficient, robust, easy to perform, and virtually unaffected by impurities due to its non-enzymatic character. DNA probes labelled with psoralen biotin are suitable for single copy gene detection in human genomic DNA when detection is performed with a streptavidin-alkaline phosphatase conjugate in combination with the enhanced dioxetane substrate LumiPhos 530™. Oligonucleotides can also be labelled with psoralen biotin by increasing the molar ratio of psoralen biotin to bases. Psoralen biotinylated oligonucleotides are suitable as both hybridization probes and PCR primers.

MATERIALS AND METHODS

DNA Probe labeling

A cDNA insert coding for the human vimentin gene[3] (1.1 kb, 50 µg/ml in TE buffer) was heat-denatured by boiling for 10 min in a water bath. After quick-chilling in an ice-water bath 1µl psoralen biotin from the RAD-FREE™ Southern Kit for Probe Labeling and Hybridization was added for every 10µl of denatured DNA solution. The DNA solution was irradiated for 1h at an intensity of 50 - 100 mW/cm2 with a 365 nm light source (RAD-FREE™ UV-lamp) in a microtiter plate on an ice bath. To remove non-incorporated psoralen biotin the labelled

Modern Applications of DNA Amplification Techniques
Edited by Lassner *et al.*, Plenum Press, New York, 1997

probe was extracted twice with water saturated n-butanol and stored on ice until hybridization. Labelled probes may be stored at -18 °C for at least 1 year. Labelling of oligonucleotides was essentially performed as described above with an increased offering ratio of psoralen biotin to bases (RAD-FREE™ Oligo-Kit).

Determination of Incorporation Extent by Competitive Immunoassay

Nucleic acids were labelled with psoralen biotin using the standard molar ratio 1 : 8 of psoralen biotin : bases. Microtiter plates were coated with anti-biotin monoclonal antibody and blocked overnight with TBS, 1% blocking reagent. Dilutions of either irradiated psoralen biotin standards without nucleic acid, or psoralen biotinylated nucleic acids were prepared in blocking solution containing 200ng/ml biotinylated alkaline phosphatase. After binding for 1h, unbound material was removed by washing 3x with TBS, 1% blocking reagent. p-nitrophenyl phosphate was added to detect bound alkaline phosphatase and incubated for 0.5h at 37 °C and the reaction stopped with NaOH/EDTA. Using a microtiter plate reader, the absorbance of all wells was measured at 405nm. The concentration of biotin in the psoralen biotinylated nucleic acids was determined from the linear portion of the standard curve.

Hybridization and Detection

Blotting and hybridization was performed according to established protocols. For detection RAD-FREE™ streptavidin alkaline phosphatase conjugate was used in combination with LumiPhos™ 530 chemiluminescent substrate sheets, providing optimum delivery of the enhanced dioxetane substrate to the reporter enzyme. Best signal-to-noise ratios were obtained using the neutral nylon membrane NYTRAN™ 0.2 μm.

RESULTS/DISCUSSION

Psoralen Biotin Labeled DNA as Hybridization Probe

Psoralen biotin is a universal nucleic acid labelling reagent, which homogeneously labels all nucleic acid exposed during the labelling reaction. However, all non-radioactive labelling methods lead to structural modifications of the DNA to be used as a probe and thus may affect the hybridization efficiency. In a series of experiments the ideal labelling conditions for preparation of sensitive hybridization probes using psoralen biotin were established (data not shown). The suitability of psoralen biotin labeled probes, using the optimized labelling conditions, is shown in figure 1. The cDNA coding for the human vimentin gene was labelled with psoralen biotin according to the standard protocol and used for detection in a Southern blot of serially diluted human genomic DNA. Hybridized probe was visualized using streptavidin alkaline phosphatase conjugate and the enhanced chemiluminescent substrate LumiPhos 530™. Even in less than 1μg of human DNA, this single copy gene can be clearly detected.

The high detection sensitivity of psoralen biotin labelled probes is most probably due to the fact, that all DNA exposed to psoralen biotin is converted to labeled probe and thus every hybridization event on the membrane can be used for signal generation upon detection. This is an important difference tomost commonly used randomly primed probes where a significant portion of unlabeled template DNA remains in the hybridization solution. The unlabeled template DNA also hybridizes, but cannot be used for signal generation upon detection.

Incorporation Extent of Psoralen Biotin

We have quantified the extent of incorporation of psoralen biotin into different substrates by competitive immunoassay. The results for the different nucleic acid substrates are summarized in Table 1. Psoralen biotin is incorporated once every 13.4 base pairs (= 27 bases) into double-

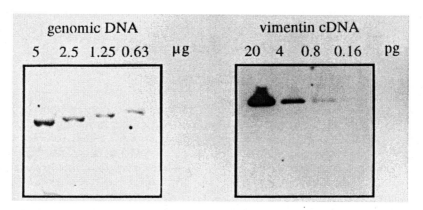

Figure 1. Single copy gene detection in human genomic DNA using a psoralen biotin labelled probe. **Left:** A dilution series of genomic DNA from the human cell line K562 (digested with BamH1) was separated on a 1 % agarose gel. The DNA was transferred to NYTRAN™ neutral nylon membrane using the TurboBlotter™ and immobilized by UV-crosslinking. The blot was prehybridized and hybridized according to established standard protocols. The EcoRI insert of the cHuVim1 plasmid[3] coding for the human vimentin gene was labelled with psoralen biotin and used at 65ng/ml in the hybridization. Detection of hybridized probe was performed using streptavidin alkaline phosphatase conjugate and the RAD-FREE™ LumiPhos 530™ chemiluminescent substrate sheets as described in the standard kit protocol. **Right:** A dilution series of the unlabelled 1.1 kb vimentin cDNA insert was blotted, hybridized and detected in parallel as a positive control. Exposure time: 2h at 37 °C.

stranded DNA and once every 30 bases into single-stranded DNA. The data suggests, that using our standard labelling conditions the affinity of psoralen biotin for DNA is roughly the same for both single- and double-stranded DNA. Previously, psoralens were primarily described as intercalators into double-stranded nucleic acids[1,2]. Increasing the salt concentration in the labelling reaction results in strongly reduced incorporation rates, while changes to the pH between 2.5 and 9.8 have no significant effect on labelling efficiency (not shown).

Labeling of RNA is also possible with a somewhat reduced incorporation rate. Presumably, the affinity of psoralen biotin to uracil is slightly lower than the affinity to thymidine, which is the primary target of psoralen in DNA. Due to the high affinity of psoralen biotin to thymidine, the incorporation rates will vary from substrate to substrate. This variation will depend on the exact base composition of the nucleic acid to be used as the probe. This is also demonstrated by the incorporation rates into a 30mer oligonucleotide (+/- 16mer TATA tail). Without the TATA tail, the incorporation rate under standard labelling conditions is one psoralen biotin per 34.1 bases. When the TATA tail is added to this oligo, the incorporation rate increases to one psoralen biotin per 12 bases, due to the higher content of thymidine.

Type of nucleic acid	1 psoralen biotin incorporated every
double-stranded DNA	13.4 base pair
single-stranded DNA	30.0 bases
oligonucleotide	34.1 bases
oligonucleotide with TATA tail (16mer)	12.0 bases
RNA	38.3 bases

Table 1. Incorporation extent of psoralen biotin into various nucleic acids

Labeling of Oligonucleotides Using Psoralen Biotin

Using the standard labelling reaction, the incorporation rate is one psoralen biotin per 34.1 bases. Statistically, this means that for a 30mer oligonucleotide part of the population will be unlabelled. When a 20mer oligonucleotide is used for psoralen biotinylation an even larger population of oligonucleotide will remain unlabelled. However, when the ratio of psoralen biotin to non-labeled bases is increased by a factor of 4, even oligonucleotides down to 16

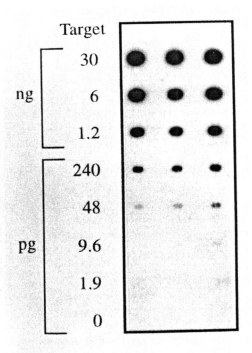

Figure 2. Dot-blot detection using psoralen biotin labelled oligonucleotides. Triplicate five-fold serial dilutions of EcoR1-digested pUC19 plasmid (30ng - 1.9pg) were blotted onto 6x SSC equilibrated 0.2 μm NYTRAN™ nylon membrane in the S&S Spotblotter (1 x 2 mm dot area). After UV-crosslinking the membrane was prehybridized for 1h in 5x SSC, 5% blocking reagent, 1% SDS at 42 °C. The oligonucleotide (GTT TTC CCA GTC ACG ACG TT) was labelled with psoralen biotin according to the RAD-FREE™ OLIGO-KIT protocol and hybridized at a probe concentration of 25ng/ml for 1h. Washing was performed twice, with 2x SSC, 0.1% SDS, for 5 min, at room temperature, and twice with preheated 0.1x SSC, 0.1% SDS, for 5min at 42 °C. After blocking, streptavidin alkaline phosphatase conjugate binding and washing the membrane was applied to RAD-FREE™ LumiPhos 530™ chemiluminescent substrate sheets and exposed at 37 °C for 3h to x-ray film.

bases, or less can be efficiently labeled and the biotinylated oligonucleotides are suitable as hybridization probes (Figure 2).

Psoralen biotin labeled oligonucleotides can also be used as primers for PCR (not shown). In this case it is preferable, that no T is present at the last or penultimate position on the 3′-end of the oligonucleotide to be labeled. Labeled primers of this type can lead to unsuccessful PCR amplification. Most probably psoralen biotin poses a steric hindrance to the polymerase, preventing elongation of the annealed primers.

TROUBLESHOOTING

Due to the non-enzymatic character of the labeling reaction, probe generation using psoralen biotin is virtually unaffected by impurities, pH, temperature deviations, and stability of enzymes or modified nucleoside triphosphates. Thus, highly biotinylated probes can be generated easily and reproducibly, leading to a very high sensitivity in non-radioactive detection[4,5,6]. According to our experience, psoralen biotin labeled probes can be used with all protocols established for ^{32}P-labelled probes, without significant deviations in hybridization efficiency. Special protocols are not necessary and the experienced user should have no problem in adopting this unique labelling system. For optimal results in probe labeling, hybridization and detection the following guidelines should be carefully observed:

1. Salt concentrations higher than 20mM in the labeling reaction lead to decreased incorporation rates. TE buffer or sterile water should be used for labeling. Precipitated DNA pellets should be washed with 70% ethanol once to remove residual salt.

2. Quickly chilling the heat denatured probe in an ice-water bath is essential because psoralen biotin contains two reactive sites on the molecule. Complementary strands in double-stranded DNA may become crosslinked. As a result the labeling efficiency is very good, but only part of the probe can hybridize. After quick-chilling the heat denatured DNA remains single-stranded for many hours when kept on ice.

3. In some cases a depression in melting temperature may be observed for oligonucleotide hybridization, due to reduced base pairing stability of psoralen biotinylated hybrids. In practice, it is advisable to start with a lower hybridization temperature e. g. 42°C, or even 25°C, and determine the optimal detection condition by increasing the hybridization temperature.

4. According to our experience, neutral nylon membranes, such as NYTRAN™, give optimal signal-to-noise ratios in dioxetane-based chemiluminescent detection. For blotting of target fragments smaller than approximately 300 bases, a pore size of 0.2μm should be used for blotting. UV-crosslinking or baking at 80 °C, is essential for immobilization of target nucleic acids.

5. The „spotty" background often observed in non-radioactive detection can in many cases be eliminated by a simple probe filtration step: The required amount of probe is mixed with 300 μl hybridization solution, filtered through a syringe filter containing a cellulose acetate membrane (Uniflo™ Plus or CELTRON™, cellulose acetate does not bind DNA/RNA or protein) and the filter rinsed once with another few hundred microliter of hybridization

solution for complete recovery of the probe. The filtration step presumably removes high molecular weight aggregates suspected to cause the background. The filtered probe is added directly to the hybridization solution.

REFERENCES

1. Cimono G. D., H.B. Gamper, S.T. Isaacs and J.E. Hearst. Psoralens as photoactive probes of nucleic acid structure and function. *Ann Rev Biochem*, 54: 1151-1193 (1985).
2. Hearst J.E., S.T. Isaacs, D. Kanne, H. Rapoport and K. Straub. The reactions of the psoralens with deoxyribonucleic acid. *Quarterly Review of Biophys*. 17: 1-44 (1984).
3. Perreau J., A. Lilienbaum, M. Vasseur and P. Paulin. Nucleotide sequence of the human vimentin gene and regulation of its transcription in tissues and cultured cells. *Gene* 62: 7-16 (1988).
4. McDonald C.L., W.E. Mahler and R.J. Fass. Revised interpretation of oxacillin MICs for *Staphylococcus epidermidis* based on *mecA* detection. *Antimicrobial Agents Chemotherapy* 39: 982-984 (1995).
5. Wang Z. and P. Dröge. Differential control of transcription-induced and overall DNA supercoiling by eukaryotic topoisomerases *in vitro*. *EMBO Journal* 15: 581-589 (1996).
6. Davidson I. and M. Malkinson. A non-radioactive method for identifying enzyme-amplified products of the reticuloendotheliosis proviral env and LTR genes using psoralen-biotin labelled probes. *J Virol Methods* 59: 113-119 (1996).

THE EFFECT OF QUANTITATIVE RATIO BETWEEN PRIMER PAIRS ON PCR PRODUCTS IN MULTI-TARGET AMPLIFICATION

Dani Bercovich, Zipi Regev, Tal Ratz and Yoram Plotsky

MIGAL - Galilee Technological Center, Kiryat Shmona 10200, Israel.

INTRODUCTION

The enormous efforts spent in studying the PCR enable us to control it through its components and conditions mainly qualitatively and partly quantitatively[1-7]. This statement is valid for single fragment amplification by PCR.

Applying the optimal conditions for two separate single fragment amplifications into multi-fragment amplification, in the same reaction, results in a different picture of the PCR product (Figure 1). This means that there is a kind of competition in the reaction between the different targets that are amplified. Because the templates that are amplified in the multiplex PCR are from different sites in the human genome (e.g. different chromosomes), there is no clear explanation to this competition.

In this study we have tried to analyze the effects of some components on multi-target PCR for a better understanding of the competitive nature which causes the difference in products between single and multi-target amplification.

MATERIALS AND METHODS

DNA preparation

DNA was extracted by phenol-chloroform method from peripheral blood collected in EDTA tube.

PCR primers

The primers were located within chromosome 21 (490bp amplicon: 5' -CTCGAG GATCCCATCCACACT-3' [21p2A], 5' GAGCCTCAGTTTTCTCCTCTG 3' [21p2B]), chromosome 1 (301bp amplicon: 5'-CCTAACTCCTGTCCCGTAACT-3' [1p5A], 5'-CTGCTCCTGGAAGGTGACAAT-3' [1p5B]), chromosome 11 (177bp amplicon: 5'-

Modern Applications of DNA Amplification Techniques
Edited by Lassner *et al.*, Plenum Press, New York, 1997

CCTAGACATTGCCCTCCAGA-3'[11p7A], 5'-ATCCCAGCAGCGTGTAGTG-3'[11p7B])
and chromosome Y (230bp amplicon: 5'-ATGAACGCATTCATCGTGTGGTC-3' [SRYA],
5'-CTGCGGGAAGCAAACTGCAATTCTT-3' [SRYB]).

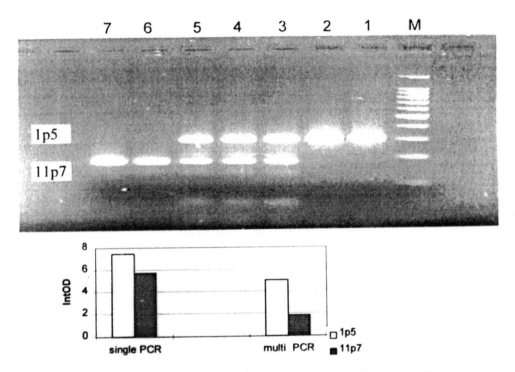

Figure 1. Comparison of PCR products between single and multi-target amplification. 9ml (from a total volume of 25 ml) of PCR product were run on a 2% ethidium bromide stained agarose. Lane **M**: 100bp DNA marker. Lanes **1** and **2**: amplicon of single targets with primer 1p5. Lanes **3, 4** and **5**: amplicon of multi-target PCR using primers 1p5 and 11p7 in the same tube. Lanes **6** and **7**: amplicon of single-target with primer 11p7. In the given plot the Y-axis represents the amount of the resulting PCR products as measured with the Eagle Eye II Image analysis.

PCR conditions

PCR was performed in a total volume of 25 µl for single and multiplex PCR with 5-10 repeats. The single components and conditions were varied during the experiments in single steps: 50 ng DNA, 2.5µl 10x reaction buffer (500mM KCl, 100mM Tris HCl pH 8.0, 15mM $MgCl_2$), 100ng of each primer, 1 unit Taq polymerase (Promega) and 0.2µM dNTPs for single-target and 1.5 units Taq polymerase and 0.3µM dNTPs for multi-target PCR.

Protocol

Denaturation 94°C for 5min followed by 25-30 cycles with 94°C for 60sec, 58°C for 60sec, 72°C for 30sec and for final extension 7min at 72°C. The amplification procedure was performed on Robocycler™ (Stratagene Inc.) with a golden gradient annealing block.

Separation and quantitation of PCR products

9μl of PCR product with 5μl of a loading-buffer were electrophoresed on a 2% agarose at 120mA for 60min. The agarose-gel was stained with ethidium bromide. The intensity of the resulting amplification bands was determined by Eagle Eye II Image analysis (Stratagene Inc.) and the quantitative analysis was performed using the RFLP-Scan software.

RESULTS

In order to get a better understanding of the nature of competition in multi-target amplification we have compared the influence of different PCR components and conditions on single-target versus multi-target amplification assays.

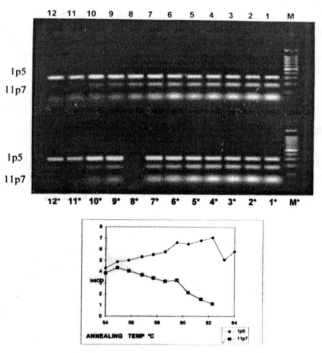

Figure 2. Gradient annealing temperature of multi-target amplification with primer pairs 1p5 and 11p7. Lane M and M*: 100bp DNA marker. Lane 1 and 1* to Lane 12 and 12*: PCR products from different annealing temperatures ranging from 54°C to 65°C, respectively. In the plot given below, the Y-axis represents the amount of the resulting PCR products as measured with the Eagle Eye II Image analysis. The X-axis gives the different annealing temperatures used in the experiment. Comparison of PCR product between single and multi-target amplification

We performed a PCR with single target amplification on two sets of primer pairs (1p5, 11p7) and a multi-target amplification with those two primer pairs in one tube (Figure 1). We used an annealing temperature of 58°C for all 3 amplification assays because this was found as the optimal annealing temperature for both single-target amplification procedures. Doing a single-target amplification the quantity of 1p5 and 11P7 products were similar, whereas in the multi-target amplification the quantity of the lower band (11p7) was significantly reduced in comparison to the upper one (1p5).

Gradient annealing temperature of multi-target amplification

We performed a multi-target amplification using the Robocycler™ gradient golden block with an annealing gradient ranging from 54°C to 64°C (Figure 2). There was an inverse relation in the quantity of PCR product between the two bands: 1p5 increased and 11p7 decreased with a rising annealing temperature. In the multi-target amplification the optimal annealing temperature for primer 1p5 was 62°C and for primer 11p7 55°C. It is important to announce that two PCR amplicons had the same optimal annealing temperature (58°C) within a single-target amplification procedure.

Different annealing time of multi-target amplification.

The rationale for this experiment was to test the possibility that the nature of the competition between the primer pairs is related to stechometric structures which cause interference to the annealing of one pair when the other pair has already been annealed.

We tested two different annealing times (60sec and 120sec) on multi-target amplification with primers 21p2 and SRY. However, the annealing time had no effect on the ratio of the two resulting amplicons (data not shown).

The effect of DNA template quantity on PCR product

We performed a single and multi-target amplification using 1ng, 10ng, 20ng, 30ng and 50ng of DNA, respectively, using primers 1p5 and 11p7.

The effect of the template amount correlated with the quantity of the resulting PCR product in both, the single and multi-target amplification assays. The amount of the template had no effect on the competitive nature of multi-target amplification. This result is in contrast with the data given by Morrison and Gannon[9].

The effect of nucleotide quantity on PCR product

Further on, we tested the effect of different quantities of nucleotides on the single-target amplification in comparison to the multi-target amplification. For primer pairs 21p2 and SRY were with two different concentrations of nucleotides (0.2mM and 0.4mM). The assays were repeated in double for the two different single-target amplification assays and in quadruple for the 4 different concentrations of nucleotides (0.1mM, 0.3mM, 0.6mM, 1.2mM) for the multi-target amplification with those two primer pairs.

Interestingly, the effect of the nucleotide concentrations was similar in the single and multi-target amplification assays, respectively, and had no effect on the competitive nature of multi-target amplification. As expected, a high concentration of nucleotides, without increasing the amount of $MgCl_2$ in the PCR buffer, resulted in an inhibition of the PCR reaction.

The effect of Taq Polymerase quantity on PCR product

For investigating the effect of different quantities of Taq Polymerase on the single-target versus the multi-target PCR procedure we have tested different concentrations of Taq polymerase: 1U and 2U in two different single-target assays using primers 21p2 and SRY and 0.5U, 1.5U, 3U, 6U, respectively, in the multi-target amplification with those two primer pairs. No clear effect of any enzyme concentration on the quantity of PCR product was demonstrable, neither in the single-target nor in the multi-target amplification procedure.

The effect of different primer quantities and primer ratios on single-target amplification assays

We tested the effect of different primer concentrations on the resulting PCR products in the single-target amplification assay using the primer pairs 21p2 and SRY in 5 different volumes (25ng, 50ng, 100ng, 200ng and 400ng). Additionally, the effect of different ratios between the two primers within the same reaction was investigated using ratios of 400ng:100ng, 200ng:100ng, 100ng:100ng, 100ng:200ng and 100ng:400ng. This procedure was repeated for the two primer pairs 21p2 and SRY. As a result we found within a certain range no significant effect of the different primer concentrations on the resulting PCR product.

The effect of primer ratio in multi-target amplification on PCR product

In a multi-target amplification assay we have tested the effect of different primer ratios on the resulting amount of the PCR product (Figure 3). For primer pairs we used 21p2 and SRY in 5 different ratios: we kept the quantity of one primer pair constant (100ng 21p2) while changing the quantity of the second primer pair (SRY) to cover the range from 1:4 to 4:1.

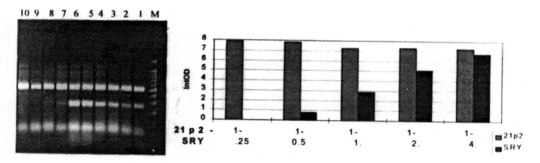

Figure 3. Different ratios of the primer pairs used in a multi-target amplification assay. For primer pairs 21p2 (upper band) and SRY (lower band) were used. Lane M: 100bp DNA marker. Lanes 1 and 2: multi-target amplification with 100ng of primer pair 21p2 and 100ng of primer pair SRY. Lanes 3, and 4: 100ng 21p2 and 200ng SRY. Lanes 5 and 6: 100ng 21p2 and 400ng SRY. Lanes 7 and 8: 100ng 21p2 and 50ng SRY. Lanes 9 and 10: 100ng 21p2 and 25ng SRY. In the plot the Y-axis represents the amount of resulting PCR product as measured with the Eagle Eye II Image analysis.

An equimolar ratio between the two primer pairs resulted in a larger amount of the PCR product for the 21p2 fragment than for the SRY fragment. Changing the ratio between primer pairs in favor of SRY resulted in decreasing the difference in quantity of product between the fragment. Using a ratio of 4:1 for SRY to 21p2 we obtained a similar quantity of product for both fragment.

To verify these results we compared in a large number of repeats (Figure 4) multi-target amplification with 1:1 primer pairs ratio versus 1:4 primer pairs ratio for 21p2 and SRY. As a control we increased in a 4-fold manner the quantity of primers in the single-target amplification assays. However, in the single-target assays there was almost no effect of varying the primer concentrations on the resulting PCR product. In contrast, in the multi-target assay the effect of increasing the quantity of primer pair SRY to 400ng, while the quantity of primer pair 21p2 was constantly 100ng, resulted in a significant effect on the ratio of PCR products.

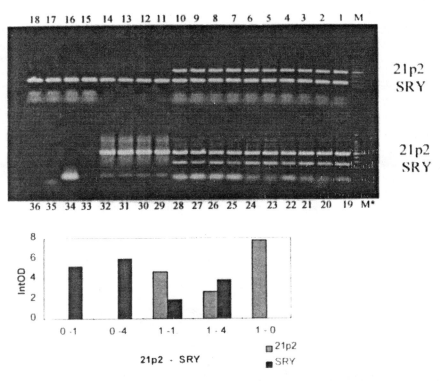

Figure 4. Multiple repeats of different primer pair ratios in multi and single-target amplification, respectively using primer pairs 21p2 and SRY. Lane M and M*: 100bp DNA marker. Lanes 1 to 10: amplicons from multi-target assays with 100ng primer pair 21p2 and 400ng primer pair SRY. Lanes 11 to 14: amplicons from single-target assays with 100ng primer SRY. Lanes 15 to 18: amplicons from single-target assays with 400ng primer SRY. Lanes 19 to 28: amplicons from multi-target assays with 100ng primer pair 21p2 - and 100ng primer pair SRY. Lanes 29 to 32: amplicons from single-target assays with 100ng primer pair 21p2. Lane 34: negative control without enzyme. Lane 35: negative control without DNA. In the plot given below, the Y-axis represents the amount of resulting PCR product as measured with the Eagle Eye II Image analysis.

DISCUSSION AND CONCLUSION

We found clear differences in the relating quantity of PCR products for the two fragments in a multi-target amplification assay in comparison to a single-target one (Figure 1).This indicates that there is a kind of competition between the different fragments. During the study we tried to analyze the nature of this competition. We came to the conclusion that the competition is not due to a shortage of any of the PCR components. The ratio between the primer pairs in multi-target amplification has a strong effect on this competition: by changing the ratio between the primer pairs it is possible to control the differences in the quantity of the resulting PCR products for the two fragments.

The important conclusion from this work is that our knowledge of the PCR reaction is still poor. There is an urgent need for a better understanding what happens in the PCR tube when doing mutiplex-target amplification. Understanding the competitive nature of multi-target amplification has great influences on the economy and simplicity of PCR technology, especially of quantitative PCR assays using internal standards.

REFERENCES

1. Christodoulou K., P. Ioannou and L. Middleton. Molecular genetic detection of Xp21 muscular dystrophy carriers in Cyprus. *Biomed Pharmacother,* 48: 355-358 (1994).
2. Jeffrey S., R. Chamberlain, A. Gibbs, E. J. Ranier and C.T. Caskey. Multiplex PCR for the diagnosis of Duchenne Muscular Dystrophy. PCR protocols. Ed. Michael A. Innis. Academic Press, INC. Harcourt Brace Jovanovich, Publishers, p272-281 (1990).
3. Apostolakos M.J., W. H. Schuermann, M. W. Frampton, M. J. Utell and J. C. Willey. Measurement of gene expression by multiplex competitive polymerase chain reaction. *Anal Biochem,* 213: 277-284 (1993).
4. Abbs S. and M. Bobrow. Analysis of quantitative PCR for the diagnosis of deletion and duplication carriers in the dystrophin gene. *J Med Genet,* 29: 191-196 (1992).
5. Palejwala V.A., R.W. Rzepka; D. Simha and M.Z. Humayun. Quantitative multiplex sequence analysis of mutational hot spots- Frequency and specificity of mutations induced by a site-specific ethenocytosine in M13 viral DNA. *Biochemistry,* 32: 4105-4111 (1993).
6. Pertl B., U. Weitgasser, S. Kopp, P. M. Kroisel, J. Sherlock and M. Adinolfi. Rapid detection of trisomies 21 and 18 and sexing by quantitative fluorescent multiplex PCR. *Hum Genet,* 98: 55-59 (1996).
7. Mansfield E.S. Diagnosis of Down syndrome and other anaploidies using quantitative polymerase chain reaction and small tandem repeat polymorphisms. *Hum Mol Genet,* 2: 43-50 (1993).
8. Eggeling F.V., M. Freytag, R. Fashold, B. Horsthemke and U. Claussen. Rapid detection of trisomy 21 by quantitative PCR. *Hum Genet,* 91:567-570 (1993).
9. Morrison C. and F. Gannon. The impact of the PCR plateau phase on quantitative PCR. *Biochim Biophys Acta,* 1219: 493-498 (1994).

PCR Quantification of Infectious Agents

QUANTITATION OF *RUBELLA* VIRUS GENOME BY QPCR AND ITS APPLICATION TO RESOLUTION FOR MECHANISM OF CONGENITAL *RUBELLA* SYNDROME

Shigetaka Katow [1] and Satoko Arai [1,2]

[1] Department of Viral Disease and Vaccine Control, National Institute of Health, Musashi-Murayama, Tokyo 208, Japan, [2]Department of Applied Biological Science, College of Bioresource Sciences, Nihon University, Fujisawa, Kanagawa 225, Japan

INTRODUCTION

Rubella virus infection of women during early stage of pregnancy often induces congenital defects in the newborn [1]. Major defects include deafness, cataracts and heart disorders and are collectively known as Congenital *Rubella* Syndrome (CRS). At every epidemic of *Rubella*, there was reoccurence of babies born with CRS and also of a considerable number of artificial abortions because of the fear of having babies with CRS [2].

To diagnose the fetal infection during the pregnancy, prenatal genome detection of the *Rubella* virus was developed in our laboratory, using three kinds of tissues of fetal origin, such as, chorionic villi, amniotic fluid and umbilical cord blood [3]. By the method, 209 cases have been diagnosed without any doubtful result [4]. From 40.0% of 100 cases of apparent infection with rash in the mother, *Rubella* virus genomes were detected. While, only 7(6.4%) out of 109 cases of inapparent infection, were genome positive. In summary, 47 out of 209 cases were proven to be genome positive. As no baby was born with CRS from 162 cases of the genome negative group, the diagnostic method is thought as a highly reliable one. From 47 cases of genome positive group, 7 cases were born beyond artificial abortion. Only two babies out of these 7 cases, had congenital disorders characteristic of CRS (induction rate of disorder: 28.6%). This does not mean that all fetuses infected *in utero* during the pregnancy would necessarily express the accurate CRS. Thus this poses the question what parameters determine the fate of the infected fetus to develop CRS. For example, one can speculate, that the dynamic viral spread in the fetus is the crucial point. When the viral growth in the fetus reaches a level that is greater than a certain critical balance, the fetus will develop CRS.

To confirm this hypothesis, a PCR-assay (QPCR) for the quantitation of the virus genome

Modern Applications of DNA Amplification Techniques
Edited by Lassner *et al.*, Plenum Press, New York, 1997

was developed. Using this method allows to resolve clinical features and CRS induction mechanism more clearly.

MATERIALS AND METHODS

Clinical specimens

All clinical specimens were collected after the informed consent of the patients or - in the case of infants - the parents of the patients. After sampling all specimens were stored at -70°C.

RNA extraction

Total RNA was extracted from the clinical specimens using an acid guanidium-thiocyanate-phenol-chloroform method [5]. After extraction RNA was dissolved in sterile RNase-free water.

PCR amplification

The PCR amplification was done using the nucleotides 8548-8916 from the E1 polypeptide coding region of the *Rubella* virus genome (nucleotide numbering of the *Rubella* virus genome was used from a sequence [6], with the addition of two nucleotides to the sequence being reported [7]). A five microliter aliquot of the extracted RNA sample was given to a 20µl reaction mixture containing 50mM Tris-HCl (pH 8.3), 75mM KCl, 3mM $MgCl_2$, 10mM dithiothreitol, 1mM of each dNTP, 200 Units of RNase inhibitor (Takara,Otsu), and 100 Units of reverse transcriptase (Superscript RT RNase H⁻, GIBCO BRL, Gaithersburg, MD), and 0.33µg of two oligonucleotides, one with the sequence 5'-CCGAGGCCCCACCGGGACTG-3' (primer 6; complementary to nucleotides 8897-8916 of the *Rubella* virus genome) and the other with the sequence 5'-GGCCTCTTACTTCAACCCTG-3' (primer 5; nucleotides 8548-8567). The reaction mixture was incubated at 37°C for 60min. An aliquot of 5µl of the reverse transcription reaction mixture was added to a 30µl PCR reaction mixture containing 1 x Taq buffer (Promega, Madison WI; 50mM KCl, 10mM Tris-HCl [pH9.0], 0.1% Triton X-100, 1.5mM $MgCl_2$), 0.2 mM of each dNTP, 1 Unit of Taq DNA Polymerase (Promega), and 3ng of two oligonucleotides primers (primer #5 and #6). The PCR reaction cycle parameters were 94°C for 1min, 55°C for 2min, and 72°C for 1.5 min for 30 cycles with an additional incubation of 72°C for 2.5 min at 30th cycle (PC-700 thermal cycler, Astec, Fukuoka). 2µl of the PCR reaction were added to a second 50µl PCR reaction mixture containing 1 x Taq buffer (Promega), 0.2mM of each dNTP, 1 Unit of Taq DNA polymerase (Promega), and 0.33µg of two oligonucleotides primers, one with the sequence 5'- TCGGGCGGGACCTGGACCTC-3' (primer 2; complementary to nucleotides 8831-8850 of the *Rubella* virus genome) and the other with the sequence 5'-GCGGCAGCTACTACAAGCAG- 3' (primer 19; nucleotides 8568-8587). The resulting DNA fragments were separated by an agarose gel electrophoresis and stained with ethidium-bromide.

Quantitation of cDNA

Quantitation of cDNA of *Rubella* virus genome was performed by nested PCR using a QPCR system 5000 (Perkin-Elmer, Norwalk CT). For quantitation of cDNA in clinical specimens, 15 cycles of the first PCR with two oligonucleotides primers 5 and 6, and 22 cycles of QPCR as the second PCR with two oligonucleotides primer 19 labeled with Tris-(2,2'-bipyridine) ruthenium (II) chelate (TBR) and primer 2 labeled with biotin [8] were used. Parameters of PCR and QPCR were the same as above-mentioned for the amplification procedure. After QPCR, biotin moiety of the amplified DNA fragments was bound with streptavidin conjugated magnetic Dynabeads followed by a washing step (Figure 1). Luminosity of activated TBR by oxidation reaction was determined with a photomultiplier at 620nm and quantitated as amount of cDNA.

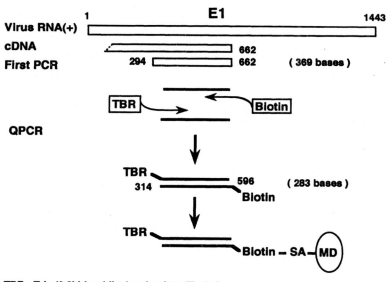

Figure 1. Procedure of quantitation of cDNA by RT-nested QPCR. Nucleotide numbering in the figure is from the first nucleotide(No.1) to the last nucleotide(No.1443) of *Rubella* virus E1 gene.

RESULTS AND DISCUSSION

Quantitation of DNA

The specificity of the oligonucleotides amplifying the E1 gene of *Rubella* virus was proven by using three different DNA templates (Table 1). There was a linear increase of luminosity from amplification cycle 10 to cycle 22. After the 23rd cycle luminosity reached a plateau value.   Below a luminosity limit of 13,000 the luminosity values linearly correlated with the amount of the DNA template. To plot the luminosity values of the clinical specimens less than 13,000, amplification cycles of the first and the second QPCR were set at 15 and 22, respectively.   Improvement of detection spectrum to the higher amount of DNA using the same procedure would be required to customize this system.

Template	*Measles* virus	*Rubella* virus	
Virus gene	NP	NS4	E1
Luminosity	9	76	>13,000

Table1. Specificity of labeled oligonucleotides primers to template cDNA. cDNAs of viral gene were amplified by a QPCR system for 25 cycles using labeled primers 19 and 2.

Infection of fetus with *Rubella* virus was a systemic infection

Rubella virus genomes were widely detected in multiple organs from the fetuses aborted artificially due to the infection of mothers during the early stage of pregnancy and also from a baby that died with severe CRS (Table 2). This argues for the fact that this infection with *Rubella* virus causes a typical systemic infection. In the cataract lens of another CRS baby high luminosity was observed (Table 2 B).

A. Artificially aborted cases due to intrauterine infection.

#[1]	Organ	Lumin.[2]	#[1]	Organ	Lumin.	#[1]	Organ	Lumin.
1	Cerebrum	27	2	Cerebrum	49	3	Cerebrum	32
	Chorionic villi-1	22		Fetal blood	11		Diaphragm	29
	Chorionic villi-2	15		Kidney	11		Fetal blood	19
	Chorionic villi-3	15		Liver	848		Heart	902
	Lens-right	22		Placenta	1,138		Lens-1	64
	Lens-left	18		Umbilical cord	83		Lens-2	13
	Skin	17					Liver	12
							Pancreas	131
							Placenta	86
							Spinal cord	5,076
							Spleen	1,153
							Testis	121

B. Case of a baby who died at 16 days of age with CRS.

Organ	Luminosity	Organ	Luminosity	Organ	Luminosity
Adrenal gland	621	Kidney	853	Spleen	527
Bladder	642	Lens	11,680	Stomach	21
Pericardial fluid	42	Cerebrum	653		
Heart	1,850	Serum	29		

Table 2. *Rubella* virus genomes detected in the various organs of the fetuses and a baby infected in utero with the virus.[1] denotes for the three different cases given in the table. [2]Lumin. denotes for luminosity

Infection of fetus with *Rubella* virus was a persistent infection

Once the virus has infected the fetus, it persists in the fetal organs during the pregnancy and even after birth. About one year later it seems to be cleared (Figure 2). In cataract lenses, however, the viruses persisted until removal of the infected tissues. If the infected tissues are not removed by an operation the viruses might persist lifelong. In our experience, virus was detectable as late as about 4 years after birth.

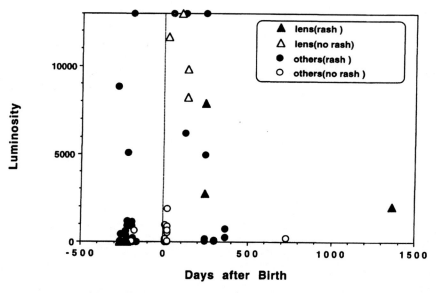

Figure 2. Quantitation of cDNA of *Rubella* virus genome in the specimens from the fetuses and infants of intrauterine infection. All samples used in the figure showed positive results in the virus genome detection by a regular nested PCR. „Others" means organs other than the lenses (adrenal gland, bladder, blood, cerebrum, cerebrospinal fluid, heart, kidney, pericardial fluid, serum, skin, spleen, stomach, throat swab, urine in the infants and amniotic fluid, blood, cerebrum, chorionic villi, diaphragm, heart, kidney, liver, pancreas, placenta, skin, spinal cord, spleen, testis, thymus, umbilical cord in the fetuses). The maximum detectable luminosity value in the QPCR system was 13,000.

IgG antibodies against *Rubella* virus were transmitted from the mother to the fetus, whilst IgM antibodies were produced in the fetus from around 20 weeks of gestation.   However, these antibodies could not eliminate the viruses from the fetal body although they had neutralizing activity when tested *in vitro* [9]. The elimination of the virus from the infected tissues is a cell mediated immunity process. At the age of one year, immunity became completely mature and thus could eliminate the virus from the body. In contrast to other organs, lenses have a quite privileged immunological situation with scarce blood circulation and lymphocyte infiltration as well. This constellation causes the possibility of a lifelong failing viral eliminination.

Viral genome amounts monitored by the luminosity assay resulting from fetuses and babies whose mothers showed typical rash were higher than those in the fetuses and babies whose mothers didn't show rash (Figure 2). In contrast to other organs, the values in the lenses had shown no difference between fetuses and babies whose mothers showed rash and those who didn't. This finding supports the previous observation that when the virus has entered the lens it grows and persists without any immunological elimination.

Pathway of virus transmission from the mother to the fetus

Plotting the the viral load against the intervals between the onset of rash in the mothers and the time of tissue sampling for the prenatal diagnosis, the transmission pathway of the virus from the mother to the fetus becomes obvious (Figure 3). The virus spreads from the mother to the chorionic villi at an interval of at least 10 days and grows there until 25 days after rash.

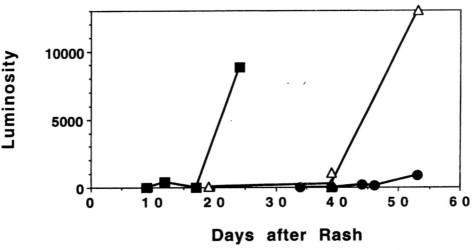

Figure 3. Transmission and time intervals (given in days) of *Rubella* virus from the infected mothers to the fetuses. Each point represents an individual sample. Closed square represents luminosity of virus genome in the chorionic villi, open triangle amniotic fluid, and closed circle umbilical cord blood. Intervals were calculated from the day of onset of rash in the mother to the date of the sampling.

Then the virus appears in the amniotic fluid as early as 19 days and in the umbilical cord blood as early as 34 days after rash, respectively. According to these data, the virus seems to be transmitted from the chorionic villi to the fetus at an interval of about 10 to 24 days. The origin of the viruses in the amniotic fluid and umbilical cord blood is thought to be secreted and circulated, respectively, from the fetus. The pathway from the chorionic villi to the fetus happens via umbilical cord blood. However, the viral genome was not detectable probably because of a low viral concentration at that moment.

TROUBLESHOOTING

Appropriate cycle for QPCR amplification

QPCR system has a limited detection capacity with a maximum luminosity of 13,000. To plot every luminosity values of the specimens proportionally to the amount of cDNA in the range of 0-13,000, appropriate cycle numbers of QPCR are necessary. Additionally, the efficacy of the first PCR-assay (before doing QPCR) should be determined in a pilot test before the next experimental step.

QPCR luminosity background

The QPCR amplification of negative controls (sample buffer instead of cDNA template) often produces high background levels. In other words, QPCR amplification with more than 22 cycles causes high background signals and reaches the maximum value (luminosity 13,000) at 30 cycles. To suppress the background levels in negative samples, amplification cycles should be restricted to less than 25 steps. Therefore it is absolutely recommended to accompany every test assay with excessive negative controls and to subtract their resulting luminosity values from those of the test samples. In the case of *Rubella* amplification in clinical specimens, 15 cycles for the first PCR assay and 22 cycles for the QPCR step result in reproducible and reliable luminosity values.

Specificity of the Luminosity

Even in the test sample, unexpected high value of luminosity appeared sometimes. In such cases, specific DNA band shoud be confirmed in an agarose gel electrophoresis. Ideally it would be better to check the purity of the amplified products by an agarose gel electrophoresis at every luminosity assay. If no DNA band could be found at the expected position in the sample with high luminosity, it will be better to discard the data and to repeat the assay.

REFERENCES

1. Gregg, N.M. Congenital cataract following German measles in the mother. *Trans Ophthalmol Soc Aust*, 3:35 (1944).

2. Katow, S. Congenital *Rubella* syndrome in Japan. Cases in the period from 1978 to 1993. *Clinic Virol*, 23:148 (1995) (text in Japanese).

3. Katow, S. Prenatal genome diagnosis of *Rubella* virus infection. in: „Proceeding of the 5th Fukuoka International Symposium of Perinatal Medicine," Maternity and Perinatal Care Unit, Kyushu University Hospital, Fukuoka 19 (1995).

4. Katow, S. Diagnosis and prevention of intrauterine infection of *Rubella*. *Obst Gyn Prac*, 44:1883 (1996) (text in Japanese).

5. Chomczynski, P. and N.Sacchi. Single-step method of RNA isolation by acid guanidinium thiocyanate-phenol-chloroform extraction. *Anal Biochem*, 162:156 (1987).

6. Frey, T.K. Molecular biology of *Rubella* virus. *Adv Vir Res*, 44:69 (1994).

7. Dominguez G., C. Wang, T. K .Frey. Sequence of the genome RNA of *Rubella* virus: evidence for genetic rearrangement during togavirus evolution. *Virol*, 177:225(1990).

8. Clementi M., S. Menzo, P. Barnarelli, A. Manzin, A. Valemsa and P. E.Varaldo. Quantitative PCR and RT-PCR in Virology. *PCR Method Applicat*, 2:191 (1993).

9. Cooper L. Z., P. R. Ziring, A. B. Ockerse, B. A. Fedun, B. Kiely, S. Krugman. *Rubella*: clinical manifestations and management. *Am J Dis Child*,118:18 (1969).

SIGNIFICANCE OF THE DETECTION OF *RUBELLA* VIRUS RNA BY NESTED PCR IN PRENATAL DIAGNOSTICS OF VIRAL INFECTIONS

Barbara Pustowoit

Institute of Virology, University Leipzig, Liebigstr. 24, 04103 Leipzig, Germany

INTRODUCTION

Rubella infection especially during the first trimester of pregnancy can cause damages to the embryonic organs and may also result in late symptoms such as endocrinopathies and others. Prenatal diagnosis of fetal infection is difficult. IgM has been detected in fetal blood samples occasionally as early as 15 weeks of pregnancy. Unfortunately blood can be withdrawn under ultrasound guidance from the umbilical vein only after 20 weeks in many fetuses. Not every infection of pregnant women with *Rubella* virus causes prenatal damage of the child and so pregnancies were often interrupted unnecessarily. With the help of molecular biological methods we should be able to diagnose a prenatal *Rubella* infection earlier than in the 20th week of pregnancy.

For a better understanding of the structure of the native rubella antigen we investigated the virus using well characterized murine monoclonal antibodies. With a panel of synthetic peptides (SPs) we defined the molecular determinants of a murine monoclonal antibody (mAb) library[1,6,15,16,20,21]. The *Rubella* virus epitopes of this mAb are demonstrated in figure 1.

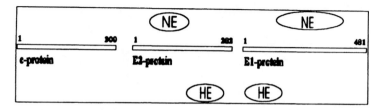

Figure 1. Functional important epitopes of the *Rubella* virus structural proteins (nucleocapsid c, E1 and E2), NE-neutralizing epitopes, HE-hemagglutinating epitopes.

The SPs were also used to investigate the human cellular immune responses [12]. In this study, we characterised the value of the SPs to evaluate T-cell and humoral responses following acute natural *Rubella* infection and vaccination.

With the knowledge of the functional properties of this antibodies our impression about the functionally important epitopes at the *Rubella* virus- hemagglutinating and neutralizing epitopes was completed (Figure 1).

In cooperation with the Department for Gynecology of University of Leipzig the prenatal diagnostic of viral infections was introduced in 1989. Using punction of free moving pupertal venal fetal blood samples were taken. To differentiate antibodies from mother and child with a special staining we were sure, that the blood samples had a fetal origin. The basics for the prenatal viral diagnostics therefore, include:

-The availibility of a highly sensitive Ultrasonar in gynecological centres to insure that the right sort of blood samples, amniotic fluid and amniotic villi can be drawn,

- The diagnostic systems must be sensitiv enough, to measure the level of fetal IgM antibodies and specific enough, to measure only clinically relevant amounts of viral antibodies and a nested *Rubella* PCR must be present.

- Most important of all is the cooperation between pregnant women, gynecologist and diagnostic laboratories to assure the right and complex interpretation of the results.

Since the introduction of *Rubella* vaccination programs, the number of CRS cases have declined remarkably. However, *Rubella* virus infections during pregnancy still occur. Therefore it is necessary to further improve diagnostics for *Rubella* infections during pregnancy as soon as possible. Patients with CRS were shown to have a decreased humoral and cellular immunity [3]. It is unknown, whether asymptomatic newborns who had experienced intrauterine infection with *Rubella* virus, differ in their antibody response from newborns with CRS. We compared both groups looking for a difference which might be a useful diagnostic criterion for CRS during the prenatal and newborn periods. The new test methods we used were the *Rubella* immunoblot and the *Rubella*-IgG-peptide-enzyme immunoassay (EIA) [13].

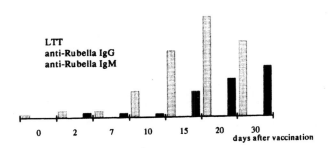

Figure 2. Time course of T-and B-cell immunity in *Rubella* infection after actual *Rubella* infection and vaccination

As expected, the study showed, that the relative proliferation rates (PPR) of wild type virus-infected patients 4 weeks after infection were higher than the PPRs of vaccinees 4 weeks after vaccination using selected synthetic peptides and RV as antigen. Surprisingly, the two synthetic peptides SP 11 and 21 caused higher PPR's in lymphocytes of vaccinees compared with wildtype virus-infected patients. This phenomenon could be due to a presentation of peptides antigen *in vitro* which might differ from processed peptide presentation which were processed naturally [19].

Another important aim of our study was to demonstrate the relation between the development of T- and B-cell immunity (Figure.2). In the time course after *Rubella* virus vaccination we were able to show that first the T-cell immunity is measurable using LTT; later on also the B-cell immunity (IgG and IgM) can be demonstrated. This result is in concordance with general knowledge of immunity development in other viral infections [17].

In immunoblot assays, produced under nonreducing conditions, we have made the interesting observation that the kinetics of the IgG response to the 3 structural proteins E1, E2 and c is different in patients as well. The response to the E1 and c protein is fast with detectable bands appearing in the first weeks following infection or vaccination. The response to the E2 protein is however delayed, with an initial response being detected only after 3 months following exposure to *Rubella* virus. This makes the use of immunoblot as supplementary test to IgM an interesting new tool in the differentiation between acute and convalescent response [17].

MATERIALS AND METHODS

Clinical specimens

The clinical specimens had been collected from the routine diagnostics. In table 1 the nested rubella PCRs being performed from the patients are given. All specimens were stored at -70°C.

RNA extraction

Total RNA was extracted from the clinical specimens using the guanidium-thiocyanate-phenol-chlorophorm method. The purified RNA was dissolved in RNase-free water.

PCR amplification

The genome of the *Rubella* virus is characterized by a single-strand plus sense RNA and is composed of only three structural proteins, named nucleocapsid (C), envelope 2 (E2) and envelope 1 (E1). For nucleic acid diagnostics, site 3 region of envelope 1 gene was reversely transcribed and amplified with the method of a nested PCR. The resulting amplification compunds of 286 base pairs. The nested *Rubella* PCR was done according to the method published by Katow [9] and coworkers.

pregnant women*	14
mother/child pair (blood/amniotic fluid)	2
children	8
fetal blood	3
umbilical cord blood	2
CRS	3
Rubella virus infected children	8
Rubella vaccinated patients	26

Table 1. List of different groups and resulting numbers (n =58) being investigated with the *Rubella* nested PCR assay. *amniotic fluid, blood, umbilical cord blood, chorionic villi biopsy)

The PCR amplification was done using nucleotides 8548-8916 in the E1 polypeptide coding region of the *Rubella* virus genome[9]. Five microliters of RNA-sample was added to a 18μl reaction mixture that contained 50mM Tris-HCl (pH 8.3), 75 mM KCl, 3 mM $MgCl_2$ and 100mM dithiothreitol, 1mM of each dNTP, 200 Units of RNase inhibitor, 100 Units of reverse transcriptase and 0.33 μg of two oligonucleotides primers. Primers given in McCarthy and coworkers[11] were used.

In figure 3 the protocol of the performed rubella nested PCR is demonstrated schematically.

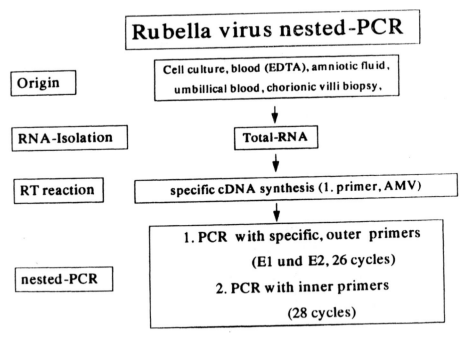

Figure 3. Protocol of the nested *Rubella* PCR

THE SIGNIFICANCE OF PCR IN *RUBELLA* DIAGNOSTICS

One of the most important roles of the immunoblot and *Rubella* nested PCR is the prenatal diagnostics.

The problem in prenatal *Rubella* diagnostics is that a considerable number of fetuses were therapeutically aborted by the fear of having babies with congenital disorders. However, in all cases there was no direct evidence of fetal *Rubella* infection. For a correct decision it is necessary and important to determine directly whether the fetus is infected or not. For this reason a *Rubella* nested PCR is important from different materials of the pregnant women using amniotic fluid, umbilical cord blood and chorionic villi biopsy.

RESULTS AND DISCUSSION

Figure 4 demonstrates a typical result of a *Rubella* nested PCR. *Rubella* virus infected cell cultures were used for controls after the first and second PCR as well. Additionally hamster ovarian cell lines with inserts of E2- and the complete 24S-gene of *Rubella* were also analysed [7,8]. This antigen can also be used in serologically tested systems [18]. To exclude inhibition factors in the amniotic fluid from the pregnant women, the *Rubella* positive control from the cell culture was added to the patient material. Figure 4 shows that all positive controls result in significant amplification product from the amniotic fluid. The amniotic fluids without viral loading were negative in all cases. We therefore can exclude an intrauterine infection of the child with a high probability.

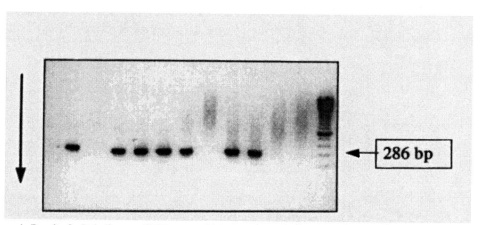

Figure 4. Result of a *Rubella* nested PCR assay with controls and amniotic fluid from an rubella virus infected pregnant women. Lane 1: positive control from the first PCR; lane 2: RT negative control; lane 3: positive control from the second PCR; lane 4: CHO cells expressing the E2-gene; lane 5: CHO cells expressing the 24S-gene); lane 6: P1 (cell culture) + positive control; lane 7: P1; lane 8: P2 + positive control (200μl); 9: P2+ positive control (500μl); lane 10: P2; lane 11: P2 (cell sediment); M: marker.

The use and significance of a nested *Rubella* PCR system is summarized in figure 5.

The PCR result can help to find *Rubella* virus genome equivalents in prenatal samples as well as in amniotic fluid, umbilical cord blood, and choriotic villi biopsy. Other suggested indications for PCR in *Rubella* infections according Carman[2] are: 1. reinfection of the mother is suspected; 2. *Rubella* appears in the second trimester of pregnancy; 3. *Rubella* appears in the first trimester, and termination of pregnancy cannot be considered on ethical causes; 4. the mother had *Rubella* before conception occured; 5. serology in the mother is unclear; 6. persistent infection is suspected.

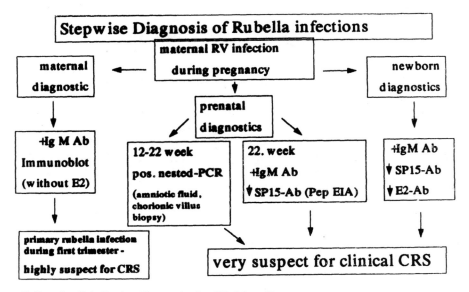

Figure 5. Complex *Rubella* virus diagnostics (modified from [19]).

TROUBLESHOOTING

Nested *Rubella* PCR is a highly sensitive method. However, there are a lot of pitfalls causing a false-positive or a false-negative results. For a correct result especially in the case of RNA is it necessary to insure a rapid isolation of RNA after sample collection, the addition of RNA-standard control to the patient's material to exclude false negative results. Nevertheless, a negative PCR result cannot exclude a *Rubella* infection. The collection of the right patient material at the right time is crucial to identify the RNA-genome equivalents correctly.

In the case of a reliable nested PCR test format, the decision for medical interventions can be based upon this new nucleic acid based approach.

However, it should be pointed that the significance of molecular biological methods in medical virology depends on several aspects:

1. Clinical evaluation, serology and molecular biology have to be considered completely;.

2. The pathogenesis of viral infections determines which material should be optimally used for establishing a diagnosis;

3. Avaliability of a specialized and well organized laboratory and short transportation.

4. Continuous validation of the establised diagnostic procedures by the use of internal controls.

REFERENCES

1. Berbers, G. Characterization of different monoclonal antibodies against *Rubella* virus in competitive elisa. *European Society for Virology*, Stockholm (1995).

2. Carman, W. F., C.Williamson, B. A. Cunliffe and A. H. Kidd. Reverse transcription and subsequent DNA amplification of *Rubella* virus RNA. *J Virol Meth*, 25: 21-30 (1989).

3. Chayne, H. H., C. A. Mauracher, A.J. Tingle and S. Gillam. Cellular and humoral responses to rubella E1, E2, and C proteins. *J Clin Microbiol*, 30: 2323-2330 (1992).

4. Chayne, H.H., P.Chong, B.Tripet, B. Brush, and S. Gillam. Localization of the virus neutralizing and hemagglutinin epitopes of e1 glycoprotein of *Rubella* virus. *J Virol*, 189: 483-492 (1993).

5. Gerna, G., M. Zannino, M. G. Revello, E. Petruzzelli and M. Dovis. Development and evaluation of a capture enzyme-linked immunosorbent assay for detection of *Rubella* immunglobulin M using monoclonal antibodies. *J Clin Microbiology*, 25: 1033-1038 (1987).

6. Green, K.Y. and P. H. Dorsett. *Rubella* virus antigens: localization of epitopes involved in hemagglutination and neutralization by using monoclonal antibodies. *J Virol*, 57: 893 (1986).

7. Hobman, T. C., M. L. Lundström, and S.Gillam. Processing and intracellular transport of *Rubella* virus structural proteins in CHO cells. *Virology*, 178: 122-133 (1990).

8. Hobman, T. C., M. L. Lundström, C. A. Mauracher, L. Woodward, S. Gillam and M. Gist-Farquar. Assembly of *Rubella* virus structural proteins into virus-like particles in transfected cells. *Virology*, 202: 575-585 (1994).

9. Katow. S. *Rubella* vaccine. In: Vaccine Handbook. Ed. by Researcher's Associates, The National Institute of Health, Maruzen, Tokyo.

10. Katow, S. Quantitation of *Rubella* virus genome by QPCR and its application to resolution for mechanism of congenital *Rubella* syndrome. Augustusburg conference of advanced science „Problems of Quantitation of Nucleic Acids by Amplification Techniques", Augustusburg, Germany, September 23-26, 1996.

11. McCarthy, M., A. Lovett, R. Kerman, A. Overstreet and J.S. Wolinsky. Immunodominant T-cell epitopes of *Rubella* virus structural proteins defined by synthetic peptides. *J Virology*, 67: 673-681 (1993).

12. Meitsch K., G.Enders, J. S. Wolinsky, R. Faber and B. Pustowoit. *Rubella*-immunoblot and rubella-peptide-EIA for the diagnosis of the congenital *Rubella* syndrome during the prenatal and newborn periods. *J Med Virol* (in press) .

13. Umino, Y., T. Sato, S. Katow, T. Matsuno, and A. Sugiura. Monoclonal antibodies directed to E1 glycoprotein of *Rubella* virus. *Arch Virol*, 83: 33-42 (1985).

14. Pustowoit, B., B. Meisegeier, H. Musielski, B. Thiel and J.Hofmann. Detection of *Rubella* IgM antibodies by IgM-capture enzyme immunoassay using monoclonal antibodies - sensitivity and

specificity of the IgM-capture ELISA. *Klin Lab*, 38: 71-279 (1992).

15. Pustowoit, B., E. Sukholutsky, J.S. Wolinsky, H.Trauer, J. Hofmann and C. Jaensch. Der Rötelnpeptid-EIA und der Rötelnimmunoblot als diagnostische Testsysteme zum Nachweis von Rötelnprimoinfektionen. Frühjahrstagung der Gesellschaft für Virologie Gießen 15.-18.3. 1995 (Poster).

16. Pustowoit, B., L. Grangeot-Keros, T.C. Hobman, and J. Hofmann. Evaluation of recombinant *Rubella*-like particles in a commercial immunoassay for detection of anti-*Rubella* IgG. *Clin Diagn Virology*, 5: 13-20.

17. Pustowoit, B., H. Trauer and J. Hofmann. Rötelnvirus. In: Diagnostische Bibliothek, Band 1 „Virusdiagnostik", Thomas Porstmann (Ed.), Blackwell Wissenschafts-Verlag Berlin-Wien.

18. Waxham, M.N. and J.S. Wolinsky. Detailed immunological analysis of structural polypeptides of *Rubella* virus using monoclonal antibodies. *Virology*, 143:153-158 (1995).

19. Wolinsky, J.S., M. McCarthy, O. Allen-Cannady, W.T. Moore, R. Jin, C. Shi-Nian, A. E. Lovett, and D. Simmons. Monoclonal antibody-defined epitope map of expressed *Rubella* virus protein domains. *J Virol*, 65: 3986-3994 (1991).

20. Zhang, T., C.A. Mauracher, L.A. Mitchell, and A. Tingle. Detection of *Rubella* virus-specific immunoglobulin g (IgG), IgM and IgM antibodies by Immunoblot Assays. *J Clin Microbiol* 30: 824-830 (1992).

QUANTITATIVE DETECTION OF HUMAN CYTOMEGALOVIRUS DNA IN CEREBROSPINAL FLUID BY POLYMERASE CHAIN REACTION

Jens-Uwe Vogel[1] and Bernard Weber[1,2]

[1]Institut für Medizinische Virologie, Universitätskliniken Frankfurt, Paul-Ehrlich-Str. 40, 60596 Frankfurt am Main, Germany. [2]Laboratoire Lieners et Hastert, 18, rue de l'Hôpital, 9244 Diekirch, Luxembourg

INTRODUCTION

Human cytomegalovirus (HCMV) is a major opportunistic pathogen in immunocompromised patients, i. e. transplant recipients and individuals with acquired immunodeficiency syndrome (AIDS). Clinical and autopsy studies have shown evidence of active HCMV infection in up to 90 % of patients with AIDS [1,2]. The spectrum of clinical manifestations attributable to HCMV infection includes retinitis, interstitial pneumonia, oesophagitis, hepatitis and colitis. Autopsy studies have revealed HCMV involvement of the central nervous system (CNS) in 20 - 30 % [3-5]. Specific HCMV neurologic syndromes that are supported by pathologic findings include subacute radiculomyelopathy, peripheral neuropathy and encephalitis [6-10]. The clinical presentation of HCMV-CNS disease is nonspecific and the role of HCMV in the development of CNS disorders is difficult to define during life. HCMV caused disorders of the CNS are often masked by the direct effects of HIV and the presence of other opportunistic pathogens [11,12]. Neurological symptoms may include meningeal signs, disorientation, short-term memory deficits, apathy, dementia, coma, seizures, or brainstem involvement [13]. Consequently, diagnosis of HCMV-CNS disease is mostly achieved by postmortem histopathologic studies [13-15]. However, early antiviral therapy has potential benefit in some of the neurologic disorders attributable to HCMV involvement of the CNS [7].

Although most patients with systemic HCMV disease have positive buffy coat cultures, virus is recovered from cerebrospinal fluid (CSF) only in about 50 % of AIDS patients with HCMV-CNS disease either by means of the conventional or shell vial culture [16,17]. Amplification of HCMV-DNA with PCR permits a non-invasive, rapid, sensitive and specific diagnosis of HCMV CNS disaease [16-19]. Overall, a high agreement between PCR results and post mortem histopathologic findings is observed.

Actually, major limitations of DNA amplification techniques, are the cumbersome nucleic

Modern Applications of DNA Amplification Techniques
Edited by Lassner *et al.*, Plenum Press, New York, 1997

acid extraction and the detection of amplificates with radioactive or non-radioactive hybridisation techniques based on southern or dot blot. Conventional hybridisation of amplification products is still the most time consuming step in the verification of the PCR amplification products and leads only to qualitative results. Alternatively, nested-PCR permits a control of the specificity of the first amplification step but is subjected to a high risk of contamination through carry over [20-22].

In the present study, a rapid extraction protocol combined with an automated non-radioactive hybridisation assay for the quantitative detection of PCR amplified HCMV-DNA from CSF was established.

MATERIALS AND METHODS

Clinical specimens

A total of 111 CSF samples were obtained from patients hospitalised at the University Clinics, Frankfurt/M., FRG, during the time interval from March until the end of November 1994 and stored frozen at -70° C within one hour after lumbar puncture. Twelve samples were collected from AIDS patients with central nervous system (CNS) disease. Five of these samples were follow-up CSF specimens from an AIDS patient with meningoencephalitis. In the first of these follow-up samples, HCMV infection of the CNS was diagnosed by virus isolation using the shell vial culture as previously described [23]. HCMV infection of the brain was confirmed by post mortem histopathologic examination. Further 24 samples were obtained from AIDS patients without clinical signs of CNS disease. Eight samples were obtained from newborns with suspicions of congenital infection. HCMV isolation from urine and DNA detection from PBMC samples were negative. The remaining samples were derived from patients suffering from CNS disease not attributable to HCMV infection. These samples were obtained as part of the evaluation for other diagnoses, including meningitis and brain neoplasms.

DNA extraction

DNA was extracted from CSF samples using a modification of the method described by Walsh et al. (1991). In a 1.5 ml tube, 100μl of CSF were added to 300μl of a solution consisting of 10% (w/v) Chelex 100 Resin (Bio Rad Laboratories, Richmond, CA) in bidestilled water. After vortexing for 10sec, the mixture was incubated at 60°C for 30min followed by vortexing for 10s. After incubation at 95°C for 10min in a thermomixer (Eppendorf, FRG), the tube cooled to room temperature. Five μl of the supernatant was carefully removed and used for direct amplification.

Primers and probe

HCMV-DNA was amplified using twenty base oligonucleotide primers, HCMV-22 and HCMV-24D which flank a 313 base pair segment of the major immediate early (MIE) sequence of HCMV-strain AD 169. Their sequences were as follows: 5'- ACT AAC CTg CAT ggg ACg Tg [position in HCMV MIE, 1878-1897]; HCMV-24D: DIG-5'- ATC TCC TCg AAA ggC TCA Tg [position in HCMV MIE, 2171-2190]) and have been selected from the Genmon data base (GBF, Braunschweig, FRG) using the Primer-Find (Fröbel, Lindau, FRG) software. One of the primers was 5′-end labelled with digoxigenin.

The amplificates were hybridized using the Enzymun-Test-DNA detection with a 5′-end biotinylated 25 base-pair probe HCMV 25SB: Bio-5′- AAg ATT AAC TCT TgC ATg TgA gCg g [position in HCMV MIE, 1926-1950]. Primers and the biotinylated probe were provided by Dianova (Hamburg, FRG).

Amplification

PCR was performed in a 50µl reaction mixture containing 50mM KCl, 10mM Tris-HCl (pH 8.3), 1.5mM MgCl$_2$ 0.1% Triton X-100, 200µM of each deoxynucleoside triphosphate, 0.5µM of each oligonucleotide primer, 5µl of extraction supernatant and 2 U of Taq DNA polymerase (Boehringer Mannheim, Mannheim, FRG). The reaction was performed in a special reagent cup with an integrated volume limitation (Sarstedt, Nuembrecht, FRG) in order to substitute oil overlaying of the reaction mixture. Samples were denatured for 1 min at 95°C and amplified in 25 cycles in a programmable thermal cycler (OmniGene, Hybaid, GB) as follows: 30sec at 94°C for denaturation, 30 sec at 48°C for annealing and 35 sec at 72°C for polymerisation. This last step was extended to 5 min on the last cycle to ensure completion of the amplified products.

Each test series included positive and negative controls. The negative control, consisting of pooled CSF samples from 20 patients without neurological disease or non-HSE encephalitis, was assayed in duplicate.

Hybridisation assay

For the hybridisation with the Enzymun DNA detection Test (Boehringer Mannheim, FRG) 40 µl of the amplification products were transferred to sample cups provided by the manufacturer and diluted with 360µl denaturation reagent. All sample cups were placed into slots of the rotation block of the ES 300 processor (Boehringer, Mannheim, FRG) and the hybridisation procedure was performed automatically by the ES 300 instrument as described elsewhere [24]. Briefly, 100µl of the denaturated amplification product and 400µl of the hybridisation solution including the biotinylated HCMV-DNA-capture probe were incubated in the streptavidin-coated tube at 37°C for 2h. After rinsing with wash solution, 500µl of anti-digoxigenin-horseradishperoxidase conjugate was added, followed by incubation for 30min. After a final washing step, 500µl of the substrate/chromogen solution was added and incubated for another 30min. The absorbance of each tube was measured at 422 nm.

Construction of plasmid *pJUV*

The 701 bp sequence within the coding region of the MIE gene of HCMV served as the basis for the generation of a control DNA sequence. The MIE of AD169 was amplified with the primer pair CMV 18 and CMV 19 according to an amplification protocol as previously described [25]. The amplification product was ligated in vector pGEM-T (Promega, Madison, WI, USA) according to manufacturer's instructions. The resulting 3702 bp plasmid, referred to as *pJUV*, was extracted from positive clones using the Qiagen purification system (Quiagen GmbH, Hilden, FRG). The amount of *pJUV* was measured by using the GeneQuant II spectro-photometer (Pharmacia, Freiburg, FRG).

DNA quantification

Different amounts of plasmid *pJUV* in 5 µl supernatant were amplified in three independent PCR reactions followed by hybridisation with the ES 300. The OD_{422} values obtained for each of the *pJUV* concentrations after hybridisation were plotted against the plasmid copy number used as PCR template to obtain a standard curve (Figure 1). This curve was constructed utilising the computer program SigmaBlot (Sigma Diagnostics, Deisenhofen, FRG),

In each PCR reaction, the supernatant of AD169 infected human foreskin fibroblasts containing 10,000 copies/5µl of HCMV-DNA served as control for the efficiency of amplification. The expected OD_{422} value of the positive control was between OD_{422} 7.00 and OD_{422} 8.00. Quantification of amplified HCMV-DNA could only be achieved if the expected value was in this range.

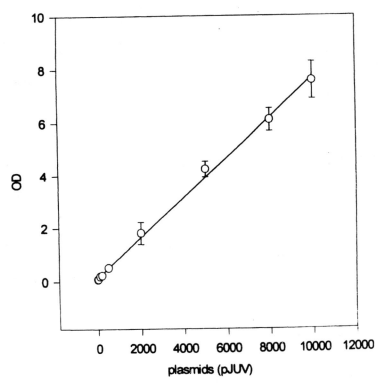

Figure 1. Linear regression analysis of standard curve obtained by plotting OD_{422} values of Enzymun-Test DNA-detection of HCMV DNA PCR amplificates against plasmid pJUV copy number

RESULTS

As shown in figure 1, the quantitative detection of three independent PCR-assays was highly reproducible. The standard deviation (SD) for the OD_{422} values of 100 plasmid copies was 0.01 whereas for 10,000 plasmid copies SD was 0.71. The results from amplifying and quantifying by hybridisation of the plasmid *pJUV* in three independent PCR-assays when analysed by linear regression, gave an $r^2 = 0.998$, verifying the reproducibility of the linear relationship between OD_{422} and HCMV MIE DNA concentration. A positive control consisting of 10,000 copies/5µl of AD169 DNA was used for the validation of each PCR-assay in order to control the efficiency of each test. Due to the high reproducibility of each assay the initially established standard curve (Figure 1) was used for all subsequent PCR runs.

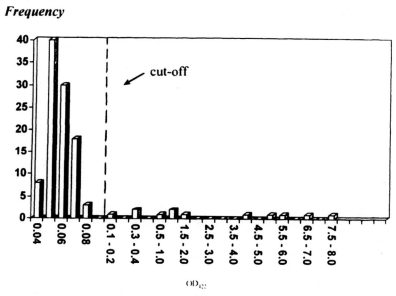

Frequency

OD_{422}

Figure 2. Distribution of OD values of Enzymun-Test DNA detection of HCMV PCR amplificates in 111 CSF samples.

Using the Enzymun-Test-DNA detection assay, a linear increase of the OD_{422}-signal in the range of 10 to 10,000 plasmid copies per 5µl used as PCR templates could be observed. Using more than 10^4 copies of *pJUV* as template in our PCR-assay a saturation effect became visible.

Of the 111 CSF samples investigated, 12 gave positive signals. Five of them were follow-up samples from an AIDS patient suffering from a meningoencephalitis. The remaining 7 positive samples were also obtained from AIDS patients with CNS disease.

A strict discrimination between positive and negative specimens was possible (Figure 2). The mean OD_{422} value of 99 negative specimens used was 0.057 ± 0.001. The cut off value was fixed at 0.09 OD_{422} with an equivocal zone of ± 10 %. Figure 2 shows the distribution of the OD_{422} values of 111 samples investigated with the PCR-assay. HCMV-DNA positive CSF samples had a virus-DNA content between 3.0 * 10^4 to 1.95 * 10^6 virus-DNA/ml.

In order to determine the sensitivity of the DNA detection assay, plasmid *pJUV* dilutions were subjected to PCR amplification and subsequent hybridisation as shown in Figure 1. The detection limit of the PCR in combination with the Enzymun-Test was 10 copies of *pJUV*/µl.

DNA extracted from mock-infected cell cultures and from cell cultures infected with herpes simplex virus type 1 and type 2 (HSV-1, HSV-2), varicella zoster virus (VZV), Epstein-Barr-virus (EBV) and human herpes virus 6 (HHV6) gave no specific amplification product in the PCR as shown in the hybridisation assay.

DISCUSSION

New hybridisation assays for the detection of PCR amplified DNA are on the verge of entering routine diagnostic laboratories since they permit fast and easy handling, shorter incubation times and yield a higher sensitivity than southern or dot blot hybridisation techniques. In comparison with other methods currently used for the diagnosis of intracerebral HCMV infection, PCR on CSF yields a higher sensitivity, is a direct and specific method and does not need invasive procedures apart from lumbar puncture. Previous studies demonstrated that PCR can detect the presence of HCMV-DNA in the CSF of patients with HIV and neurologic abnormalities [26,27].

The DNA extraction protocol using Chelex 100 instead of the traditional cumbersome phenol/chloroform or proteinase K extraction permits a rapid and efficient DNA extraction as shown previously for the detection of HSV DNA in CSF [24,28]. With the Chelex extraction procedure, potentially inhibiting blood factors are efficiently eliminated [29,30]. Furthermore the few preparation steps reduce the risk of contamination which is particularly important during the extraction procedure [20,31].

Although from a theoretical point of view nested-PCR presents a higher sensitivity than one-stage PCR, nested-PCR is connected with a higher risk of contamination through carry over.

HCMV DNA detection in 30 samples can be performed within 3.5 hours - a diagnostic result of cerebrospinal fluid with pre-treatment, amplification and hybridisation within 6 hours. In contrast to previous non commercial hybridisation assays [32] the Enzymun test permits an easy and rapid detection of amplified DNA.

Quantification methods based on co-amplification of target DNA and the same modified target sequence as internal standard have the advantages of detecting potential inhibitory effects present in the samples [32]. Whereas the competitive PCR represents the method of choice for the quantitative detection of HCMV-DNA from PBMC samples, due to the competition with the internal standard there occurs a reduced sensitivity. This might limit its usefulness for the diagnosis of HCMV encephalitis. Since virus is recovered from cerebrospinal fluid (CSF) only in about 50 % of AIDS patients with HCMV CNS disease either by means of the conventional or shell vial culture [16], HCMV-DNA is probably present at relatively low copy numbers in CSF. Therefore, it is of importance to use a highly sensitive PCR assay. Combining a one stage PCR with the Enzymun-Test-DNA detection assay, a high sensitivity and reproducibility were achieved.

Quantitative detection of HCMV-DNA in PBMC represents an useful marker for the monitoring of high risk patients. High viral DNA levels appear to correlate with the severity of HCMV disease. In contrast, relatively low amounts of HCMV-DNA may be detected in seropositive patients with latent HCMV infection or in asymptomatic patients with reactivation [25]. Although quantitative detection is not of critical importance for the diagnosis of HCMV CNS disease, quantification of viral DNA may prove as a useful marker for the antiviral therapy, since ganciclovir or foscarnet appear to be not always effective in the treatment of HCMV encephalitis.

Quantitative PCR as described in this paper also offers an easy and reliable method for the evaluation of new antiviral substances. A further advantage of our method is the cost-effectiveness since only one sample dilution and one positive control are required for quantification of viral DNA. Usually, serial target/internal standard DNA dilutions are necessary for quantification with the competitive PCR. End-point sample dilution methods are even more cost intensive and are in fact semi- quantitative assays, which allow only comparison of the relative amounts of target sequences in clinical samples.

TROUBLESHOOTING

The main source of trouble was contamination by carry-over. Contamination could however be avoided by good laboratory practice and the physical separation of the activities involved in the performance of PCR. Contamination prevention could be further improved by incorporating AmpErase (uracil N-glycosylase, UNG) which hydrolyses selectively dUTPs integrated into the amplicons before each new run. Inhibition due to incomplete seeding of the pellet or inappropriate removal of the supernatant was the most common reason for false negative results. Malfunctions of the ES processor occurred from time to time.

REFERENCES

1. Reichert, C. M., T. J. O_Leary and D. L. Levens. Autopsy pathology in the acquired immune deficiency syndrome. *Am J Pathol*, 12: 357 - 382 (1983).
2. Quinnan, G. V. jr., H. Masur and A. H. Rook Herpesvirus infections in the acquired immune deficiency syndrome. *JAMA*, 252: 72 - 77 (1984).
3. Snider, W. D., D. M. Simpson, S. Nielson, J. W. M. Gold, C. E. Metroka and J. B. Posner. Neurologic complications of acquired immune deficiency syndrome: analysis of 50 patients. *Ann Neurol*, 14: 403 - 18 (1983).
4. Petito, C. K., E.-S. Cho and W. Lehmann. Neuropathology of acquired immunodeficiency syndrome (AIDS): an autopsy review. *J Neuropathol Exp Neurol*, 45: 635 - 646 (1986).
5. Wiley, C. A. and J. A. Nelson. Role of human immunodefiency virus and cytomegalovirus in AIDS encephalitis. *Am J Pathol*, 133: 1: 73 - 81 (1988).
6. Wiley, C. A., R. D. Schrier and F. J. Denaro. Localization of cytomegalovirus proteins and genome during fulminant central system infection in an AIDS patient. *J Neuropathol Exp Neurol*, 45: 127 - 139 (1986).
7. de Gans, J., G. Tiessens, P. Portegies, J. A. Tutuarima and D. Troost. Predominance of polymorphonuclear leukocytes in cerebrospinal fluid of AIDS patients with cytomegalovirus polyradiculomyelitis *J Acquir Immune Defic Syndr*, 3: 1155 -8 (1990).
8. Said, G., C. Lacroix and P. Chemouilli. Cytomegalovirus neuropathy in acquired immunodeficiency syndrome: a clinical and pathological study. *Ann Neurol*, 29: 139- 146 (1991).
9. Edwards, R. H., R. Messing and R. R. McKendall. Cytomegalovirus meningoencephalitis in a homosexual man with Kaposi´s sarcoma: isolation of CMV from CSF cells. *Neurology*, 35: 560 - 562 (1985).
10. Fuller, G. N., R. J. Guiloff, F. Scaravilli and J. N. Harcourt-Webster. Combined HIV-CMV

encephalitis present with brainstem signs. *J Neurol Neurosurg Psychiatry,* 52: 975 - 979 (1989).

11. Navia, B. A., E. S. Cho, C. K. Petito and R. W. Price. The AIDS dementia complex: II. Neuropathology. *Ann Neurol,* 19: 525 - 535 (1986).

12. Ho, D.D., T. R. Rota and R. T. Schooley. Isolation of HTLV-III from cerebrospinal fluid and neural tissues of patients with neurologic syndromes related to the acquired immunodeficiency syndrome. *New Engl J Med,* 313: 1493 - 1497 (1985).

13. Morgello, S., E. S. Cho, S. Nielsen, O. Devinsky and C. Petito. Cytomegalovirus encephalitis in patients with acquired immunodeficiency syndrome: an autopsy study of 30 cases and a review of literature. *Hum Pathol,* 18: 289 - 97 (1987).

14. Vinters, H. V., M. K. Kwod and H.W. Ho. Cytomegalovirus in the nervous system of patients with the acquired immune deficiency syndrome. *Brain,* 112: 245 - 268 (1989).

15. Post, M. J. D., G. T. Hensley, L. B. Moskowitz and M. Fischl. Cytomegalic inclusion virus encephalitis in patients with AIDS: CT, clinical and pathologic correlation. *Am J Radiol,* 146: 1229 - 1234 (1986).

16. Gozlan, J., J. M. Salord, E. Roullet, M. Baudrimont, F. Caburet, O. Picard, M. C. Meyohas, C. Duvivier, C. Jacomet and J. C. Petit. Rapid detection of cytomegalovirus DNA in cerebrospinal fluid of AIDS patients with neurologic disorders. *J Infect Dis,* 166: 1416-1421 (1992).

17. Wolf, D.G. and S.A. Spector. Diagnosis of human cytomegalovirus central nervous system disease in AIDS patients by DNA amplification from cerebrospinal fluid. *J Infect Dis,* 166: 1412-5 (1992).

18. Atkins, J. T., G. J. Demmler, W. D. Williamson, J. M. McDonald, A. S. Istas and G. J. Buffone. Polymerase chain reaction to detect cytomegalovirus DNA in the cerebrospinal fluid of neonates with congenital infection. *J Infect Dis,* 169: 1334-1337 (1994).

19. Clifford, D. B., R. S. Buller, S. Mohammed, L. Robison, and G. A. Storch. Use of polymerase chain reaction to demonstrate cytomegalovirus DNA in CSF of patients with human immunodeficiency virus infection. *Neurology,* 43: 75-79 (1993).

20. Clewley, J. P. The polymerase chain reaction, a review of the practical limitations for human immunodeficiency virus diagnosis. *J Virol Meth,* 25: 179 - 188 (1989).

21. Cimino, G. D., K. C. Metchette, J. W. Tessman, J. E. Hearst and S.T. Isaaca. Post PCR sterilization: a method to control carryover contamination for the poymerase chain reaction. *Nucleic Acids Res,* 19: 99 - 107 (1990).

22. Gretch, D. R., J. J. Wilson, R. L. Carithers, C. de la Rosa, J. H. Han and L. Corey, L. Detection of hepatitis C virus RNA: Comparison of one-stage polymerase chain reaction (PCR) with nested-set PCR. *J Clin Microbiol,* 31: 289 - 291 (1993).

23. Schacherer, C., W. Braun, G. Bauer and H.W. Doerr. Detection of cytomegalievirus in bronchial lavage and urine using a monoclonal antibody to an HCMV early nuclear protein. *Infection,* 16: 288 - 292 (1988).

24. Sakrauski, A., B. Weber, H. H. Kessler, K. Pierer and H. W. Doerr. Comparison of two hybridization assays for the rapid detection of PCR amplified HSV genome sequences from cerebrospinal fluid. *J Virol Meth,* 50: 175 - 184 (1994).

25. Weber, B., U. Nestler, W. Ernst, H. Rabenau, J. Braner, A. Birkenbach, E.H. Scheuermann, W. Schoeppe and H. W. Doerr. Low correlation of human cytomegalovirus DNA amplification by polymerase chain reaction with cytomegalovirus disease in organ transplant recipients. *J Med Virol,* 43: 187-193 (1994).

26. Berman, S. M. and R. C. Kim. The development of cytomegalovirus encephalitis in AIDS patients receiving ganciclovir. *Am J Med,* 66: 415-419 (1994).

27. Cinque, P., L. Vago, M. Brytting, A. Castagna, A. Accordini, V. A. Sundqvist, N. Zanchetta, A.

D'Arminio Monforte, B. Wahren, A. Lazzarini, and A. Linde. Cytomegalovirus infection of the central nervous system in patients with AIDS: Diagnosis by DNA amplification from cerebrospinal fluid. *J Infect Dis,* 166: 1408-1411 (1992).

28. Kessler, H.H., K. Pierer, B. Weber, A. Sakrauski, B. Santner, D. Stuenzner, E. Gergely, and E. Marth. Detection of herpes simplex virus DNA from cerebrospinal fluid by PCR and a rapid, nonradioactive technique. *J Clin Microbiol,* 32: 1881-1886 (1994).

29. Singer-Sam, E., F. X. Heinz and C. Kunz. Use of chelex to improve the PCR signal from a small number of cells. *Amplification,* 5: 11 (1993).

31. Kwok, S. Procedures to minimize PCR-product carry-over, p142 - 145. In: PCR Protocols. A Guide to Methods and Application. M. A. Innis, D. H. Gelfand, J. J. Sninsky and T. J. White (Ed.), Academic Press, San Diego, CA, USA (1990).

32. Zipeto, D., F. Baldanti, D. Zella, M. Furione, A. Cavicchini, G. Milanesi, G. Gerna. Quantification of human cytomegalovirus DNA in peripheral blood polymorphonuclear leukocytes of immunocompromised patients by the polymerase chain reaction. *J Virol Meth,* 44: 45-56 (1993).

30. Walsh, P.S., D.A. Metzger and R. Higuchi. Chelex 100 as a medium for simple extraction of DNA for PCR-based typing from forensic material. *BioTechniques,* 10: 505-513 (1991).

QUALITY CONTROL AND EXTERNAL QUALITY ASSESSMENT SCHEMES FOR THE DIAGNOSTIC USE OF PCR IN MICRO-BIOLOGICAL LABORATORIES - EUROPEAN TRIALS ON HEPATITIS B VIRUS AND CYTOMEGALOVIRUS

Jurjen Schirm

Regional Public Health Laboratory, Groningen, The Netherlands

INTRODUCTION

In microbiological diagnostic laboratories increasing amounts of routine nucleic acid amplification tests, mostly PCR tests, are being performed. In these laboratories the danger of obtaining false positive or false negative PCR-results is relatively large: the same PCR test types are often performed in large quantities several times a week, whereas the test results are usually needed fairly quickly; in addition, many types of clinical specimens may contain inhibitors of the PCR-reaction, which may not sufficiently be removed by the rapid, relatively simple, nucleic acid extraction methods. Since the outcome of a diagnostic laboratory test may be crucial for a proper patient care the quality control of PCR testing in such laboratories is extremely important.

In relation to quality control different definitions are being used: *Quality Assurance* encompasses all measures taken to ensure the reliability of investigations and the resulting laboratory reports. *Internal Quality Control* assesses whether media, reagents and equipment are performing within specifications, and whether the performance of the individual laboratory is sufficiently reproducible. *External Quality Assessment* compares the performances of different laboratories testing the same specimens.

INTERNAL QUALITY CONTROL OF AMPLIFICATION REACTIONS

Laboratory design and code of practice[1,2]

In each PCR laboratory several internal quality control measures should be used. The personnel should be well-trained, should not work under pressure and should follow a code of practice. Each stage of the PCR should be performed in a separate working area, preferably in a separate room. Reagents and stock solutions should be prepared in the „clean room" room,

Modern Applications of DNA Amplification Techniques
Edited by Lassner *et al.*, Plenum Press, New York, 1997

which should have an outward air flow. Nothing should enter this room that has been in either of the other two PCR laboratories. Sample preparation (including nucleic acid extraction) and template addition should be performed in a second room, whereas the third room is exclusively used for amplification and post-PCR analyses.

Ideally, the latter room should have an inward airflow. Direct connecting doors between the three laboratories should be avoided. Each working area should have dedicated equipment, and also dedicated labcoats and reagents should be used. In order to limit carry over plugged pipette tips or positive displacement pipettes should be used, whereas tubes with push-fit caps should be avoided. Further procedures include the centrifugation of tubes before opening, the aliquoting of all solutions, the use of „master" PCR mixes, and the use of proper negative and positive controls (see below).

Causes of false reactions

All stages of diagnostic PCRs are potential sources of false test results (Figure 1). False negative results may be caused by poor specimen collection, improper nucleic acid extraction, inhibition of the PCR by clinical specimens, failing thermocyclers, or sub-optimal detection and/or hybridization procedures. In theory also strain variations may result in false negative results, when amplified sequences coincide with variable regions in the microorganism studied. Mutations in the primer regions may prevent the PCR reaction to take place altogether. Mutations in the regions selected for probe hybridization may make it impossible to detect and/or identify the PCR products. False positive results may arise when PCR products (amplimers) from previous tests, or native DNA from other tubes or other experiments, are accidently introduced into the test tubes. Potential ways of spreading these „contaminations" are reagents, equipment, personnel and aerosols.

A very efficient, and popular, method to neutralize the adverse effects of accidently introduced amplimers is the use of the so called UNG protocol[3]. This protocol makes use of the enzyme uracil-N-glycosylase (UNG), which is able to degrade DNA sequences - but not RNA sequences - containing uracil instead of thymidine. In short the UNG protocol works as follows: when dTTP in the PCR master mix is routinely replaced by dUTP the resulting PCR products, the amplimers, will be viable to degradation by UNG. When in subsequent identical PCR experiments the reaction mixtures are pretreated with UNG (1-10 min at room temperature) all „old" amplimers will be eliminated, whereas the native DNA from the clinical specimens will not be affected. During the first step of the actual PCR process, heating at 95°C, UNG is irreversibly inactivated. As a result the newly synthesized amplimers will not be degraded and thus only „new" amplimers can be detected.

An alternative for the UNG protocol is the use of isopsoralen compounds. Amplimers containing isopsoralen compounds can be cross-linked by UV irradiation before they are removed from the tubes for analysis. Whereas cross-linked DNA cannot serve as a template for subsequent amplification reactions, it usually does not affect the analysis of the amplimers. In practice the „pre-PCR sterilization" UNG-protocol is much more widely used than the „post-PCR sterilization" isopsoralen procedure. It is important to realize that none of these methods protects against contamination by native DNA. Therefore the use of three separate rooms, and the inclusion of various control reactions, remains essential.

Control Reactions

Both positive and negative controls should be added at various stages of the procedure (Figure 1). Two different categories of control reactions can be distinguished: <u>test controls</u>, which are needed to check whether the test procedure and the measures against contaminations have been performed properly, and <u>sample controls</u>, which are used to check the suitability of the individual clinical samples, or the nucleic acid extracts from it, for testing in the PCR.

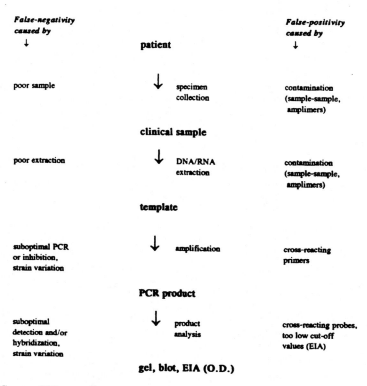

Figure 1. Causes of false reactions

Positive Test Controls

Each PCR experiment should contain positive controls in order to check the performance of the PCR step itself. Usually, different dilutions of control DNA are included after the extraction of the other samples, just prior to the amplification step. As a result the day to day sensitivity of the PCR step can be monitored quite easily. A second type of positive test control is usually included before DNA extraction, in order to test the efficiency of the extraction procedure itself. This positive control should consist of a relatively low concentration of the microorganism to be tested, and should contain no inhibitory substances. In order to <u>prevent contamination positive controls should always be added last.</u>

Negative Test Controls

Negative controls are crucial for monitoring whether all the measures against contamination have worked properly. Ideally, negative controls should be added at all different stages of the test procedure (Fig 1). However, negative controls included in the earliest stage, just after specimen collection, are most important, since these negative controls can unmask contamination irrespective of the stage where the contamination takes place. Therefore, at least two negative controls should be added in this stage, ideally at least one for every five test samples. In most laboratories additional negative controls are included after the nucleic acid extraction step, just prior to the amplification step.

A special type of negative test control is the so called „UNG control", which examines whether the UNG is still active at the concentration used. When the enzyme activity of UNG is too low, the addition of small amounts of previously synthesized (U-containing) amplimers will result in positive UNG controls, whereas the presence of active UNG will produce negative UNG controls. UNG controls should be performed periodically and, in addition, when the numbers of positive test results increase unexpectedly.

Sample Controls

Sample controls (always positive controls) are used to discover inhibitory substances in individual samples, to test the efficiency of the nucleic acid extraction procedure from each sample and, sometimes, to recognize poor specimen collection.

The use of „*specimen collection controls*" is only possible with specimens that should contain cells. Then an additional PCR on a normal cellular gene, for instance a globin gene, can be performed: the globin gene PCR should always be positive. However, the outcome of this type of control experiment will also be affected by the efficiency of the extraction procedure and/or the possible presence of inhibitors. Therefore, a negative result of a globin gene PCR will be difficult to interpret without other control experiments. Moreover, for specimens without cells, for instance cerobrospinal fluids or serum samples, PCRs on globin genes or other household genes are not a possible option.

The possible presence of PCR inhibitors in individual nucleic acid extracts is usually tested by adding external control DNA to each extract just prior to the amplification step („*inhibition controls*"). In addition, the efficiency of individual nucleic acid extractions can be tested by adding external control DNA or low numbers of the complete microorganisms (usually a laboratory strain) to each clinical specimen („*extraction controls*"). *Inhibition controls* and *extraction controls* can be added in two different ways: when the nucleotide sequence of the added control DNA is exactly the same as the DNA sequence studied the control DNA must be added to a parallel test tube, which is subsequently tested using the normal protocol; however, when DNA controls with slightly different nucleotide sequences - within the amplimer region - are available, the external control DNA can be coamplified in the same test tube, and yet distinguished from the native DNA by hybridization[4]. When the *extraction controls* are prepared in the latter way a diagnostic PCR test can, relatively easily, be adapted into a quantitative PCR assay. The need to use *inhibition controls* has been widely recognized, especially when „difficult" types of samples, such as sputum samples[4,5] are tested. As a consequence commercial firms also started to include *inhibition controls* in their PCR kits.

Specificity Controls

Notwithstanding the use of all the measures and controls described above the specificity of PCR products should always be confirmed. The conventional way of checking the identity of PCR products, visualization and size determination on agarose gels, is not reliable enough for diagnostic work. Determination of the exact size of the PCR products may be difficult, especially when non-specific bands with approximately the same sizes are present. Moreover, PCR products of the right size may still originate from unexpected cross-reacting targets. Therefore the specificity of the PCR products should be confirmed with additional or alternative methods which are, in a way, always based on the detection of specific nucleotide sequences. In practice most of these methods are not only more specific but also more sensitive than the simple analysis on gels. Presently, the following confirmation methods are available to most routine diagnostic laboratories:

Restriction enzyme analysis of the PCR products on gels. Provided that proper enzymes are available this type of analysis is easy to perform[6]. However, since the detection step is still on gels the sensitivity of the method is limited.

New amplification with a different specific primer set. This has, among others, been described for the detection of papillomaviruses[7]: after screening clinical specimens with „general" primers reacting with all types of papillomaviruses, positive specimens were retested with papillomavirus type-specific primers. When both primer sets gave the right bands on gels the presence of papillomavirus had been confirmed and, in this particular case, typed simultaneously. Since this method, as it is described here, also relies on detection on gels the sensitivity will not be optimal. Therefore the authors of the mentioned paper extended their analyses by using Southern blots and specific hybridizations, which did not only increase the sensitivity but also, even more, the specificity.

Nested PCR is in fact a special version of method described above. With this method the initial PCR products are reamplified using a second set of primers which are internal to the first pair. With this strategy positive test results can only be obtained when both sets of primers annealed in a specific way. Since this approach, in theory, also increases the sensitivity considerably, the second round PCR (i.e. the confirmation step) is usually performed on all samples, without intermediate analyses of the first round products. Final analysis of nested PCR is often performed on gels, without additional hybridization. A disadvantage of the use of nested PCR is that, according to the most frequently used laboratory protocols, the test tubes must be opened after the first series of amplifications in order to add the specific primers for the second series. At this point the danger of spreading contamination is relatively large.

Nucleic acid hybridization with a specific DNA probe With this most popular type of checking the specificity of PCR products, the detection and confirmation steps are usually combined: not the amplimers but the nucleic acid hybrids are detected. For the detection of the hybrids most laboratories perform Southern blots, dot-blots[4,5], or EIA-type assays[8], often using digoxigenin (DIG) labelled probes and enzyme labelled antibodies against DIG (Figure 2). These three methods have comparable sensitivities. Southern blot hybridization has the highest specificity since it additionally reveals the size of the amplimers. Dot-blots and, especially, EIA formats are easier to perform. Consequently the EIA format is rapidly becoming the most popular tool for the analysis of large series of diagnostic PCRs. Some of the presently available commercial PCR kits are also based on the EIA principle.

Unexpected Results

Even when a positive PCR result is supported by the proper reaction controls and specificity controls one cannot entirely rule out that the detected sequence was not present in the original clinical specimen. In addition, occasionally positive results may be found in specimens from patients with later evidence of an entirely different diagnosis than originally anticipated. For these reasons, many laboratories choose to reconfirm all positive results by repeating the complete procedure, starting again with the clinical specimen. Sometimes such retests are negative indeed, indicating that the original positive result may have been wrong, or perhaps borderline.

If an unexpected negative result is found, for instance when a patient persistently carrying a microorganism becomes negative without a reasonable explanation, there is also reason for concern. Even when all the positive controls of a test run were positive one should still be aware of the possibility that the thermocycler does not produce the same temperature profiles in all tubes. Therefore it may be useful to repeat unexpected negative results with another thermocycler, or using a different part of the same cycler. In addition, the temperature profiles of the thermocyclers should be checked regularly.

EXTERNAL QUALITY ASSESSMENT

During the last few years different large scale *External Quality Assessment* schemes for the diagnostic use of PCR in microbiological laboratories have been organized. Since the preparation, distribution and evaluation of these types of schemes is quite labour intensive and costly they were usually supported in one way or another by international organizations, such as the European expert group on viral hepatitis (EUROHEP), the European Group for Rapid Viral Diagnosis (EGRVD), or the World Health Organization (WHO). In all schemes coded test panels were used, whereas the results were evaluated anonymously. The results obtained with of a number of recently investigated panels are briefly summarized below (see also Table 1).

Hepatitis B Virus DNA

In 1993 EUROHEP and the EGRVD collaborated in the organization of an *External Quality Assessment* scheme for the detection of Hepatitis B virus DNA[9]. The panel consisted of twelve undiluted serum samples, 7 HBV-DNA positives and 5 HBV-DNA negatives, and two dilution series containing 3,000,000 to 0.03 particles per ml. Dilution series were considered to be correct if one or more of the least dilute samples were found to be positive and all the higher dilutions were found to be negative, so for instance: + / + / + / - / - is a correct dilution series, but + / + / - / + / - is a wrong one.

Of the 43 participating laboratories only 19 (44%) were able to detect all 7 positive samples correctly, of whom only 10 (23%) produced faultless results with all the samples. In addition 9 laboratories (21%) missed the weak positive samples only, which led to the conclusion that the quality of 19 laboratories (44%) was sufficient (assuming that missing only some weak positive samples is less unacceptable than producing false positive results). Consequently, no less than 24 (56%) of the laboratories reported either false-positive results (15 laboratories,

35%) or incorrect dilution series or both. 60% of the laboratories had faultless results with both dilution series.

The detection limit in the dilution series from participants with good performances varied from 300 to 30,000 particles per ml. Whereas the different participants used a broad range of laboratory procedures - they all used „in-house" amplification methods - good performances were not associated with particular extraction procedures or amplification protocols.

Southern blot:

| PCR product | → | sepa-ration on gels | → | nylon filter | ← | DIG labelled probe | ← | enzyme labelled anti-DIG | ← | colour development on paper |

Dot-blot:

| PCR product | → | nylon filter | ← | DIG labelled probe | ← | enzyme labelled anti-DIG | ← | colour development on paper |

Enzyme Immuno Assay (EIA): (alternative EIA-formats are also possible)

| biotin-ylated PCR product | → | strept-avidin coated microtiter well | ← | DIG labelled probe | ← | enzyme labelled anti-DIG | ← | colour development in solution |

Figure 2. Detection of PCR products after hybridization with specific DNA probes

CMV-DNA Panel

In 1995 a CMV-DNA panel was prepared and distributed under the auspices of the EGRVD[10]. The design of the panel was similar to the HBV-DNA panel described above: 10 undiluted blood samples (2 positive, 2 weak positive, 6 negative), 10 undiluted serum samples (2 positive, 2 weak positive, 6 negative), and a dilution series of 8 samples containing 3,000,000 to 0.1 DNA copies per ml.

56 laboratories submitted a total of 63 data sets, all produced with „in-house" developed PCR assays. Of these data sets only 12 (19%) reported faultless results with all the samples. In addition 15 datasets (24%) missed the weak positive samples only, which indicates that 27

(43%) of the datasets were of reasonable quality. Surprisingly, datasets obtained with nested PCR produced better results - also less false positives - than datasets obtained with the usual single-round PCR. The reason for this observation is not known. Possibly, laboratories using nested PCR take the measures against contamination more seriously. Also in this study, the majority of the datasets (57%) showed either false-positive results (30%) or incorrect dilution series or both. Faultless results with both dilution series were reported by 71% of the laboratories. The reported detection limits in the dilution series were in the range of 100 to 10,000 copies per ml.

HCV-RNA Reference Panels

Both in 1992 and in 1994 EUROHEP distributed HCV-RNA reference panels, each consisting of 4 HCV positive plasma samples, 6 HCV negative plasma samples and two dilution series. The 1992 panel was tested by 31 different laboratories[11], and the 1994 panel by 86 laboratories[12]. In the 1994 study altogether 136 datasets were produced, 99 by using „in house" developed PCR assays, 28 using commercial assays, and 9 using other amplification methods. The results of both studies were remarkably similar: whereas in 1992 only 5/31 laboratories (16%) produced faultless results in all samples, the corresponding figure in 1994 was identical: 22/136 datasets (16%). The numbers of datasets missing only weak positive samples were 7 (23%) and 39 (29%) respectively. Consequently the percentages of datasets with acceptable results increased from 39% in 1992 to only 45% in 1994. In adddition, false-positive results were reported in 29% and 21% of the datasets respectively. Faultless results with both dilution series were reported in 48% of the datasets in 1992 and 52% in 1994. The reason for the disappointing degree of improvement is unknown. Since both studies were evaluated anonymously it is not known whether or not individual laboratories improved their performance.

Results	HCV-RNA 1992[11] (n=31)	HBV-DNA 1993[9] (n=43)	HCV-RNA 1994[12] (n=136)	CMV-DNA 1995[10] (n=63)	*Mycobacterium tuberculosis* DNA 1995[13] (n=30)
Faultless results	16 %	23 %	16 %	19 %	17 %
Total sufficient quality	39 %	44 %	45 %	43 %	37 %
False-positive results	29 %	35 %	21 %	30 %	57 %

Table 1. Summary of the results of five recent *External Quality Assessment* schemes.

Surprisingly, the performances of the commercial tests in the 1994 study were only slightly better than those of the „in-house" tests, though the difference was not statistically significant: whereas 18/28 (64%) of the commercial test sites (all using Amplicor, ROCHE) reported acceptable results, 5 (28%) of them reported false positive results in the undiluted samples.

Mycobacterium tuberculosis DNA Panel

A WHO panel for the detection of *Mycobacterium tuberculosis* DNA was distributed and tested in 1995[13]. For the preparation of this panel artificial „clinical" specimens were constructed by mixing clump-free *Mycobacterium bovis* (belonging to the *M. tuberculosis* complex) with negative sputum samples. The panel consisted of 10 sputum samples without mycobacterial cells, 5 samples with 100 mycobacterial cells (per 0.2 ml), and 5 samples with 1000 mycobacterial cells (per 0.2 ml). The panel was tested by 30 different laboratories, 22 using „in-house" PCR methods and 8 using commercial amplification kits. Faultless results were produced by only 5 laboratories (17%), whereas another 6 laboratories (20%) missed only some of the low positive samples. No less than 17 laboratories (57%) produced false positive results. Further evaluation showed that reliability was not associated with the use of any particular laboratory method, including different commercial methods.

CONCLUSIONS

Different *External Quality Assessment* studies have shown that it is difficult to perform microbiological diagnostic PCR tests reliably. Due to sensitivity problems 48-64% of the participating laboratories with acceptable performances reported false- negative results. False-positivity due to contamination was reported by 29-57% of the laboratories and is therefore a major problem. Whether or not all laboratories which performed badly in the studies do rely on their diagnostic PCR in their daily routine work is not known, but most probably not. Obviously the laboratories with problems should do something about it. Clearly, the use of commercial diagnostic PCR kits will not automatically solve all the problems.

Modern commercial automated PCR machines may be helpful, but these instruments have not been tested yet in *External Quality Assessment* schemes. Well-trained personnel is crucial, whereas proper „in-house" control measures and the performance of various control reactions also remain absolute necessities. In addition, there is a need for defined reference panels, which can be regularly tested as part of the *Internal Quality Control* system of each laboratory. When all these actions have been taken regular participation in *External Quality Assessment* schemes will reveal whether the complete *Internal Quality Control* system is still functioning properly.

ACKNOWLEDGEMENTS

I thank Dr. Wim Quint (Delft, the Netherlands) for helpful discussions and for allowing me to use some of his unpublished results on CMV quality control.

REFERENCES

1. Kitchin P. A. and J.S. Bootman. Quality control of the polymerase chain reaction. *Rev Med Virol,* 3: 107-114 (1993).
2. Dragon E.A., J.P. Spadoro and R. Madej, Quality control of polymerase chain reaction, *in:* „Diagnostic Molecular Microbiology; Principles and Applications", D.H. Persing, T.F. Smith, F.C. Tenover and T.J. White, ed., American Society for Microbiology, Washington, D.C., 160-168 (1993).
3. Persing D.H. Polymerase chain reaction: trenches to benches (Minireview). *J Clin Microbiol,* 29: 1281-1285 (1991).
4. Noordhoek G.T., J.A. Kaan, S. Mulder, H. Wilke and A.H.J. Kolk. Application of the polymerase chain reaction in a routine microbiology laboratory for detection of *Mycobacterium tuberculosis* in clinical samples. *J Clin Path* 48: 810-814 (1995).
5. Schirm J., L.A.B. Oostendorp and J.G. Mulder. Comparison of Amplicor, in-house PCR, and conventional culture for detection of *Mycobacterium tuberculosis* in clinical samples. *J Clin Microbiol,* 33: 3221-3224 (1995).
6. Vogels W.H.M., P.C. van Voorst Vader and F.P. Schröder. *Chlamydia trachomatis* infection in a high-risk population: comparison of polymerase chain reaction and cell culture for diagnosis and follow-up. *J Clin Microbiol,* 31: 1103-1107 (1993).
7. van den Brule A.J.C., C.J.L.M. Meijer, V. Bakels, P. Kenemans and J.M.M. Walboomers. Rapid detection of human papillomavirus in cervical scrapes by combined general primer-mediated and type-specific polymerase chain reaction. *J Clin Microbiol,* 28: 2739-2743 (1990).
8. Ossewaarde J.M., M. Rieffe, G.J. van Doornum, C.J.M. Henquet and A.M. van Loon. Detection of amplified *Chlamydia trachomatis* DNA using a microtiter plate-based enzyme immunoassay. *Eur J Clin Microbiol Infec Dis,* 13: 732-740 (1994).
9. Quint W.G.V., R.A. Heijtink, J. Schirm, W.H. Gerlich, H.G.M. Niesters. Reliability of hepatitis B virus DNA detection. *J Clin Microbiol,* 33: 225-228 (1995).
10. Pauw W., H.G.M. Niesters, W. van Dijk, J. Schirm, G. Gerna, H. Johansson and W.V.G. Quint. CMV-DNA Quality Assurance Program, manuscript in preparation.(preliminary report *in:* European Group for Rapid Viral Diagnosis, Newsletter, June 1995).
11. Zaaijer H.L., H.T.M. Cuypers, H.W. Reesink, I.N. Winkel, G. Gerken and P.N. Lelie. Reliability of polymerase chain reaction for detection of hepatitis C virus. *Lancet,* 341: 722-724 (1993).
12. Damen M., H.T.M. Cuypers, H.L. Zaaijer, H.W. Reesink, W.P. Schaasberg, W.H. Gerlich, H.G.M. Niesters and P.N. Lelie. International collaborative study on the second EUROHEP HCV-RNA reference panel. *J Virol Methods,* 58; 175-185 (1996).
13. Noordhoek G.T., J.D.A van Embden and A.H.J. Kolk. Reliability of nucleic acid amplification for the detection of *Mycobacterium tuberculosis:* an international collaborative quality control study among 30 laboratories. *J Clin Microbiol,* 34: 2522-2525 (1996).

CONTRIBUTORS

Albert, Jan
Department of Clinical Virology, Swedish Institute for Infectious Disease Control, Karolinska Institute, 105 21 Stockholm, Sweden

Arai, Satoko
Department of Viral Disease and Vaccine Control, National Institute of Health, Musashi-Murayama, Tokyo 208, Japan. Phone: -425-611960, FAX: -425-610771 and Department of Applied Biological Science, College of Bioresource Sciences, Nihon University, Fujisawa, Kanagawa 225, Japan

Beator, Jens
Schleicher and Schüll Inc., Hahnestraße 3, 37586 Dassel, Germany. Phone: -5561-791449, FAX: -5561-791536

Bercovich, Bani
MIGAL Galilee Technology Center, PO Box 90000, Rosh Pina 12100, Israel. FAX. -6-944980

Birikh, Klara
Arbeitsgruppe Eckstein, Max-Planck-Institut für Experimentelle Medizin, Hermann-Rein-Straße 3, 37075 Göttinge, Germany. Phone: -551-3899326, FAX: -551-3899388

Burghoff, Robert L.
Schleicher and Schüll Inc., PO Box 2012, Keene, NH 03431, USA. Phone: -603-3523810, FAX: -603-3573627

Dörsam, Volker
Institut für Biochemie, FB15, Justus-Liebig-Universität Giessen, Heinrich-Buff-Ring 58, 35392 Giessen, Germany

Droessler, Karl
Institut für Zoologie, Universität Leipzig, Talstraße 33, 04103 Leipzig, Germany

Hahn, Meinhard
Institut für Biochemie, FB15, Justus-Liebig-Universität Giessen, Heinrich-Buff-Ring 58, 35392 Giessen, Germany

Harvey, Michael A.
Schleicher and Schüll Inc., PO Box 2012, Keene, NH 03431, USA. Phone: -603-3523810, FAX: -603-3573627

Katow, Shigetaka
Department of Viral Disease and Vaccine Control, National Institute of Health, Musashi-Murayama, Tokyo 208, Japan. Phone: -425-611960, FAX: -425-610771

Ladusch, Mathias
Institut für Zoologie, Universität Leipzig, Talstraße 33, 04103 Leipzig, Germany

Lassner, Dirk
Institut für Klinische Chemie und Pathobiochemie, Arbeitsbereich Gendiagnostik, Universität Leipzig, Liebigstraße 16, 04103 Leipzig, Germany. Phone: -341-9722273, FAX: -341-9605126

Legoux, Pascale
Sanofi Recherche, Centre de Labége, Labége Innopole Voie No. 1, BP 137, 31676 Labége CEDEX, France. Phone: -6100-4000, FAX: -6100-4001

Lundeberg, Joakim
Department of Biochemistry and Biotechnology, KTH, Royal Institute of Technology, Teknikringen 30, 100 44 Stockholm, Sweden. Phone: -8-7908757, FAX: -8-245452

Milde, Jörg
Institut für Klinische Chemie und Pathobiochemie, Arbeitsbereich Gendiagnostik, Universität Leipzig, Liebigstraße 16, 04103 Leipzig, Germany. Phone: -341-9722273, FAX: -341-9605126

Minty, Adrian J.
Sanofi Recherche, Centre de Labége, Labége Innopole Voie No. 1, BP 137, 31676 Labége CEDEX, France. Phone: -6100-4000, FAX: -6100-4001

Mix, Eilhard
Neurologische Klinik und Poliklinik, Neurobiologisches Forschungslabor, Universität Rostock, Gehlsheimerstraße 20, 18147 Rostock, Germany. Phone-381-4949540, FAX: -381-4949542

Pingoud, Alfred
Institut für Biochemie, FB15, Justus-Liebig-Universität Giessen, Heinrich-Buff-Ring 58, 35392 Giessen, Germany

Plotsky, Yoram
MIGAL Galilee Technology Center, PO Box 90000, Rosh Pina 12100, Israel. FAX. -6-944980

Pustowoit, Barbara
Institut für Klinische Chemie und Pathobiochemie, Arbeitsbereich Gendiagnostik, Universität Leipzig, Liebigstraße 16, 04103 Leipzig, Germany. Phone: -341-9722273, FAX: -341-9605126

Ratz, Tal
MIGAL Galilee Technology Center, PO Box 90000, Rosh Pina 12100, Israel. FAX. -6-944980

Regev, Zipi
MIGAL Galilee Technology Center, PO Box 90000, Rosh Pina 12100, Israel. FAX. -6-944980

Remke, Harald
Institut für Klinische Chemie und Pathobiochemie, Arbeitsbereich Gendiagnostik, Universität Leipzig, Liebigstraße 16, 04103 Leipzig, Germany. Phone: -341-9722273, FAX: -341-9605126

Rolfs, Arndt
Neurologische Klinik und Poliklinik, Neurobiologisches Forschungslabor, Universität Rostock, Gehlsheimerstraße 20, 18147 Rostock, Germany. Phone-381-4949540, FAX: -381-4949542

Schirm, Jurjen
Regional Public Health Laboratory, van Kentwich Verschuurlaan 92, 9721 SW Groningen, The Netherlands. Phone: -50-5215100, FAX: -50-5271488

Shire, David
Sanofi Recherche, Centre de Labége, Labége Innopole Voie No. 1, BP 137, 31676 Labége CEDEX, France. Phone: -6100-4000, FAX: -6100-4001

Stark, Malin
Teknikringen 30, 100 44 Stockholm, Sweden. Phone: -46-8-7908757, FAX: -46-8-245452

Syvänen, Ann-Christine
Department of Human Molecular Genetics, National Public Health Institute, Mannerheimintie 166, 00300 Helsinki, Finland. Phone: -90-47441, FAX. -90-4744408

Uhlén, Mathias
Department of Biochemistry and Biotechnology, KTH, Royal Institute of Technology, Teknikringen 30, 100 44 Stockholm, Sweden. Phone: -46-8-7908757, FAX: -46-8-245452

Uhlmann, Volker
Neurologische Klinik und Poliklinik, Neurobiologisches Forschungslabor, Universität Rostock, Gehlsheimerstraße 20, 18147 Rostock, Germany. Phone-381-4949540, FAX: -381-4949542

Vener, Tanya
Department of Biochemistry and Biotechnology, KTH, Royal Institute of Technology, Teknikringen 30, 100 44 Stockholm, Sweden. Phone: -46-8-7908757, FAX: -46-8-245452

Vogel, Jens-Uwe
Institut für Medizinische Virologie, Universitätskliniken Frankfurt, Paul-Ehrlich-Straße 40, 60596 Frankfurt am Main, Germany

Vörtler, Claus Stefan
Arbeitsgruppe Eckstein, Max-Planck-Institut für Experimentelle Medizin, Hermann-Rein-

Straße 3, 37075 Göttingen, Germany. Phone: -551-3899326, FAX: -551-3899388

Weber, Bernard
Institut für Medizinische Virologie, Universitätskliniken Frankfurt, Paul-Ehrlich-Straße 40,
60596 Frankfurt am Main, Germany and Laboratoire Lieners et Hastert, 18, rue de l'Hospital,
9244 Diekirch, Luxembourg

Suppliers of Specialist Items

Advanced Magnetics, Inc., 61 Mooney Street, Cambridge, MA 02138, USA

Aldrich Chemical Co., 940 West Saint Paul Avenue, Milwaukee, WI 53223, USA

Ambion Inc., 2130 Woodward St.#200, Austin, Texas 78744-1832, USA

Amersham Buchler GmbH & Co.KG, Gieselweg 1, 38110 Braunschweig, Germany; Phone: -49-5307-2060; Fax: -49-5307-206237

Amicon Div. W.R. Grace & Co.-Conn, 72 Cherry Hill Drive, Beverly, MA, 01915, USA; Upper Mill, Stonehause, Gloucester, GL 10 2BJ, UK

Amicon GmbH; Salinger Feld 32, 58454 Witten (Ruhr), Germany; Phone: -49-2302-801996; Fax: -49-2302-800905

Anachem Ltd, 20 Charles Street, Luton, Bedfordshire, LU2 0EB, UK; Phone:-44-582-456666; Fax: -44-582-391768

Applied Biosystems, Division of Perkin-Elmer, 850 Lincoln Center Drive, Foster City, CA 94404, USA

Applied Biosystems, Brunnenweg 13, 64331 Weiterstadt, Germany; Phone: -49-6150-1010; Fax: -49-6150-101101

Appligene, Handschuhsheimer Landstr. 22, 69120 Heidelberg, Germany; Phone: -49-6221-409058; Fax: -49-6221-411062

Bachofer GmbH, PO Box 7058, D-7410 Reutlingen, Germany

Beckman Instruments, Inc., 2500 Harbour Boulevard, PO Box 3100, Fullerton, CA 92634, USA; Frankfurter Ring 115, 80807 München, Germany;

Becton Dickinson, Immunocytometry Systems, 2350 Qume Dr., San Jose, CA 95131, USA; Tullastr. 8-12, 69126 Heidelberg, Germany

Biochrom KG, Leonorenstr. 2-6, 12247 Berlin, Germany

Bio-Med, Gesellschaft für Biotechnologie, Schloß Ditfurth, 97531 Theres, Germany

Biometra, PO Box 157, Maidstone, Kent, ME14 2AT, UK

Bio-Rad Laboratories, Alfred Nobel Drive, Hercules, CA 94547, USA

Bio-Rad Laboratories GmbH, Heidemannstr. 164, 80939 München, Germany; Phone: -49-89-318840; Fax: -49-89-31884100

BIOS Laboratories, 5 Science Park, New Haven, CT 06511, USA

Biotecx Labs Inc., 6023 South Loop East, Houston, Texas 77033, USA

Boehringer Mannheim GmbH, Biochemica, Sandhofer Str. 116, POB 310120, 68305 Mannheim 31, Germany; Phone: -49-621-7590; Fax: -49-621-7598509

British Biotechnology Products Ltd, 4-10 The Quadrant, Barton Lane, Abingdon, Oxfordshire, OX14 3YS, UK

Burdick & Jackson, Division of Baxter Diagnostics Inc., 1953 South Harvey Street, Muskegon, MI 49442, USA

Cambridge Research Biochemicals Inc., Fairfax Research Room 205, Wilmington, DE 19897, USA

Cangene Corporation, 3403 American Drive, Mississauga, Ontario L4V1T4, Canada

Clontech Laboratories Inc., 4030 Fabian Way, Palo Alto, California 94303-4607, USA

Clontech Laboratories Inc., via ITC Biotechnology GmbH, Postfach 103026, 69020 Heidelberg, Germany; Tel: 06221-303907; Fax: 06221-303511

Collaborative Research, Inc., 2 Oak Park, Bedford, MA 01730, USA; Phone: -1-617-2750004; Fax: -1-617-2750043

Costar, Nucleopore GmbH, Vor dem Kreuzberg 22, 72070 Tübingen, Germany; Phone: -49-7071-94970; Fax: -49-7071-949716

Cruachem Inc., 45150 Business Ct., 550 Sterling, VA 22170-6702, USA; Phone: -1-703-6893390; Fax: -1-703-6893392

Diagen GmbH, Niederheider Str. 3, 40589 Düsseldorf, Germany

Drummond Scientific Company, 500 Parkway, Box 700, Broomall, PA 19008, USA

Dunn Labortechnik GmbH, Postfach 1104, 5464 Asbach, Germany

DuPont de Nemours GmbH, Du-Pont-Str. 1, 61352 Bad Homburg, Germany; Phone: -49-6172-872552; Biotechnology Systems Division, BRML, G-50986 Wilmington, DE 19898, USA

Dynal Inc., 475 North Station Plaza, Great Neck, NY 11021, USA

Dynal UK Ltd, Station House, 26 Grove Street, New Ferry, Wirral, Merseyside, L62 5A2, UK Deutsche Dynal GmbH, Schaartor 1, 20459 Hamburg, Germany; Phone: -49-40-366811; Fax: -49-40-366040

Eastman Kodak, Acorn Field, Knowsley Industrial Park North, Liverpool L33 72X, UK

Enzo Diagnostics; 60 Executive Blvd., Farmingdale, NY 11735, USA; Phone: -1-516-4968080; Fax: -1-516- 6947501

Epicentre Technologies, 1202 Ann Street, Madison, WI 53713, USA

Eppendorf-Netheler-Hinz GmbH, Barkhausenweg 1, 22339 Hamburg, Germany; Phone: -49-40-538010; Fax: -49-40-53801556; 45635 Northport Loop East, Fremont, CA 94538, USA

Ericomp, 6044 Cornerstone Court West, Suite E, San Diego, California 92121, USA

Finnzymes OY, PO Box 148, 02201 Espoo, Finland; Phone: -35-80-42080 77; Fax: -35-80-420 8653

Fisher Scientific, 711 Forbes Ave., Pittsburgh, PA, USA

Flowgen Instruments Ltd, Broad Oak Enterprise Village, Broad Road, Sittingbourne, Kent ME9 8BR, UK

Fluka Chemical Corp., 980 S. Second Street, Ronkonkoma, NY 11779-7238, USA

FMC BioProducts, via Biozym Diagnostik GmbH, POB 180; 31833 Hessisch-Oldendorf, Germany; Phone: -49-5152-2075

FMC BioProducts (Europe), Risingevej 1, DK-2665 Vallensbaek Strand, Denmark

Fotodyne Inc., 16700 West Victor Road, New Berlin, WI 53151-4131, USA; Phone: -1-414-7857000; Fax: -1-414-7857013

Gibco/BRL, PO Box 68, Grand Island, NY 14072-0068, USA

Gibco/BRL, Life Technologies GmbH, Dieselstr. 5, 76344 Eggenstein, Germany; Phone: -49-721-78040; Fax: -49-721-780499

Gilson France SA, via Abimed Analysen Technik, Raiffeisenstr. 3, 40764 Langenfeld, Germany; Phone: -49-2173-89050; Fax: -49-2173-890577

Glen Research Corporation, 44901 FalconPlace, Sterling, VA 22170, USA

Grant Instruments (Cambridge) Ltd., Techn. Büro Deutschland, Goethestr. 5, 65203 Wiesbaden, Germany; Phone: -49-6121-600666; Fax: -49-6121-4186746

Greiner GmbH, Postfach 1162, D-7743 Frickenhausen, Germany

Heraeus Instruments, Inc., 111-A Corporate Blvd., S. Plainfield, NJ07080, USA

Hoefer Scientific Instruments, Serva, Carl-Benz-Str. 7, 69115 Heidelberg, Germany; Phone: -49-130-7047; Fax: -49-6221-502188

Hoffmann-La Roche AG, Diagnostica, Emil-Barell-Str. 1, 79639 Grenzach-Wyhlen, Germany

Hybaid Omnigene, 111-113 Waldegrave Road, Teddington, Middlesex TWII 8LL, UK; Phone: -44-181-6141000; Fax: -44-181-977 0170

Hyperion Inc. 14100 S.W. 136th Street, Miami, FL 33186, USA; Phone: -1-305-2383020; Fax: -1-305-2327375

ICN Biochemicals Ltd., Unit 18, Thame Park Business Centre, Wenman Road, Thame, Oxon OX9 3XA, UK

Invitrogen Corporation, 3985 B Sorrento Valley Building, San Diego, CA 92121, USA

Kreatech Diagnostics, PO Box 12756, 1100 AT Amsterdam, The Netherlands; Phone: -31- 20-6919181; Fax: -31-20-6963531

Lark Sequencing Technologies Inc., 9545 Katy Freeway, Suite 200, Houston TX 77024-9870, USA

Life Technologies Ltd, Trident House, Renfrew Road, Paisley PA3 4EF, UK

Merck, 6100 Darmstadt 1, Postfach 4119, Germany

Midwest Scientific; PO Box 458; 228 Meramec Station Road, Valley Park, MO 63088, USA; Phone: -1-314-225-9997; Fax: -1-314-225-2087

MilliGen/Biosearch, Division of Millipore, 1986 Middlesex Turnpike, Burlington, MA 01803, USA

Millipore Ltd, Hauptstr. 87, 65760 Eschborn, Germany

MJ Research Inc., via Biozym Diagnostik, Hameln, Germany; Phone: -49-5151-7311; Fax: -49-5151-7313

National Biosciences, 3650 Annapolis Lane, Plymouth, MN 55447, USA

New England Biolabs/C.P. Laboratories, PO Box 22, Bishop's Stortford, Hertfordshire, CM23 3DH, UK

New England Biolabs, Postfach 2750, 6231 Schwalbach/Taunus, Germany; Tel: -49-6196-3031; Fax: -49-6196-83639

Novabiochem Ltd., 3 Heathcoat, Building, Highfields Science Park, University Blvd., Nottingham NG7 2 QJ, UK; Phone: -44-602-430840; Fax: -44-602-430951

Novagen, 565 Science Dr., Madison, Wi 53711, USA

Nunc, Inc., 2000 North Aurora Road, Naperville, IL 60563-1796, USA; Phone: -1-708-9835700; Fax: -1-708-4162556

Organon Teknika, Boseind 15, 5281 RM Boxtel, The Netherlands

Peninsula Laboratories, Inc., 611 Taylor Way, Belmont, CA 94002, USA; Neckarstaden 10, 69117 Heidelberg, Germany

Perkin-Elmer Holding GmbH, Bahnhofstrasse 30, 8011 Vaterstetten, Munich, Germany

Pharmacia LKB Biotechnology Inc., 800 Centennial Avenue, PO Box 1327, Piscataway, NJ 08855-1327, USA

Pharmacia Biotech GmbH, Munzinger Str. 9, 79111 Freiburg, Germany; Phone: -49-761-49030; Fax: -49-761-4903306

Phenix Research Products, 3540 Arden Road, Hayward, CA 94545, USA; Phone: -1-800-7670665; Fax: -1-510-2642030

Pierce, Pierce Europe BV, Box 1512, 3260 BA Ous-Beijerland, The Netherlands

Polaroid Corp., Technical Imaging Products, 575 Technology Square, Cambridge, MA 02139, USA; Ashley Road, St. Albans, Herts, AL1 5PR, UK

Polaroid GmbH, Sprendlinger Landstr. 109, 63069 Offenbach, Germany; Phone: -49-69-84041; Fax: -49-69-8404321

Polygen GmbH, Karlstr. 10, 63225 Langen, Germany

Polyscientific Corp., 70 Cleveland Avenue, Bay Shore, NY 11706, USA; Phone: -1-516-586-

0400; Fax: -1-516-2540618

Promega Ltd, Delta House, Enterprise Road, Chilworth Research Centre, Southampton, SO1 7NS, UK

Promega Corporation, Serva, Carl-Benz-Str. 7, 69115 Heidelberg, Germany; Phone: -49-130-7047; Fax: -49-6221-502188

QIAGEN GmbH, Max-Volmer-Str. 4, 40724 Hilden, Germany; Phone: -49-2103-892230; Fax: -49-2103-892222

Research Genetics Inc., 2130 Memorial Parkway, Hunteville, Alabama 35801, USA

Sarstedt Ltd. 68 Boston Road, Beaumont Leys, Leicester LE4 1AW, UK

Savant Instruments Inc.,110-103 Bi-Country-Blvd., Framingdale, NY 11735, USA

Schleicher and Schüll Inc., 10 Optical Avenue, Keene, NH 03431, USA; PO Box 4, 37586 Dassel, Germany; Phone: -49-5561-7910; Fax: -49-5561-791533

Serva Feinbiochemica GmbH & Co., Carl-Benz-Str. 7, 69115 Heidelberg, Germany;

Sigma Chemical Company (UK), Fancy Road, Poole, Dorset, BH17 7NH, UK

Stratagene, 11099 North Torrey Pines Road, La Jolla, CA 92037, USA; PO Box 105466, 69121 Heidelberg, Germany; Phone: -49-130-840911; Fax: -49-130-400639

Stratagene Ltd, Cambridge Innovation Centre, Cambridge Science Park, Milton Road, Cambridge, CB4 4GF, UK

TaKaRa Biomedicals, Takara Shuzo Co., Ltd., Biomedical Group, Otsu, Shiga, Japan; Phone: -81-775-437247; Fax: -81-775-439254

Techne Duxford, distributed by Thermo-Dux GmbH, Ferdinand-Friedrich-Str. 5, 97877 Wertheim, Germany; Phone: -49-9342-880188; Fax: -49-9342-880191

Techne Incorporated, 3700 Brunswick Pike, Princeton, NJ 08540, USA

Tri-Continent Scientific Inc., 12555 Loma Rica Drive, Grass Valley, CA 95945, USA

Tropix Incorporated, 47 Wiggins Avenue, Bedford, Massachusetts 01730, USA

United States Biochemical Corp., PO Box 22400, Cleveland, OH 44122, USA

Vector Laboratories, 16 Wulfrie Square, Bretton, Peterborough, PE3 8 RF, UK

Index